Probability and Statistics for STEM

A Course in One Semester

Synthesis Lectures on Mathematics and Statistics

Editor
Steven G. Krantz, *Washington University, St. Louis*

A Gyrovector Space Approach to Hyperbolic Geometry
Abraham Albert Ungar
2008

Probability and Statistics for STEM: A Course in One Semester
E.N. Barron and J.G. Del Greco

ISBN: 978-3-031-01299-0   paperback
ISBN: 978-3-031-02427-6   ebook
ISBN: 978-3-031-00273-1   hardcover

DOI 10.1007/978-3-031-02427-6

A Publication in the Springer series
SYNTHESIS LECTURES ON MATHEMATICS AND STATISTICS

Lecture #33
Series Editor: Steven G. Krantz, *Washington University, St. Louis*
Series ISSN
Print 1938-1743   Electronic 1938-1751

Probability and Statistics for STEM

A Course in One Semester

E.N. Barron
Loyola University, Chicago

J.G. Del Greco
Loyola University, Chicago

SYNTHESIS LECTURES ON MATHEMATICS AND STATISTICS #33

ABSTRACT

One of the most important subjects for all engineers and scientists is probability and statistics. This book presents the basics of the essential topics in probability and statistics from a rigorous standpoint. The basics of probability underlying all statistics is presented first and then we cover the essential topics in statistics, confidence intervals, hypothesis testing, and linear regression. This book is suitable for any engineer or scientist who is comfortable with calculus and is meant to be covered in a one-semester format.

KEYWORDS

probability, random variables, sample distribution, confidence intervals, prediction intervals, hypothesis testing, linear regression

Dedicated to Christina

– E.N. Barron

For Jim

– J.G. Del Greco

Contents

Preface

Every student anticipating a career in science and technology will require at least a working knowledge of probability and statistics, either for use in their own work, or to understand the techniques, procedures, and conclusions contained in scholarly publications and technical reports. Probability and statistics has always been and will continue to be a significant component of the curricula of mathematics and engineering science majors, and these two subjects have become increasingly important in areas that have not traditionally included them in their undergraduate courses of study like biology, chemistry, physics, and economics. Over the last couple of decades, methods originating in probability and statistics have found numerous applications in a wide spectrum of scientific disciplines, and so it is necessary to at least acquaint prospective professionals and researchers working in these areas with the fundamentals of these important subjects. Unfortunately, there is little time to devote to the study of probability and statistics in a science and engineering curriculum that is typically replete with required courses. What should be a comprehensive two-semester course in probability and statistics has to be, out of necessity, reduced to a single-semester course. This book is an attempt to provide a text that addresses both rigor and conciseness of the main topics for undergraduate probability and statistics.

It is intended that this book be used in a one-semester course in probability and statistics for students who have completed two semesters of calculus. It is our goal that readers gain an understanding of the reasons and assumptions used in deriving various statistical conclusions. The presentation of the topics in the book is intended to be at an intermediate, sophomore, or junior level. Most two-semester courses present the subject at a higher level of detail and address a wider range of topics. On the other hand, most one-semester, lower-level courses do not present any derivations of statistical formulas and provide only limited reasoning motivating the results. This book is meant to bridge the gap and present a concise but mathematically rigorous introduction to all of the essential topics in a first course in probability and statistics. If you are looking for a book which contains all the nooks and crannys of probability or statistics, this book is not for you. If you plan on becoming (or already are) a practicing scientist or engineer, this book will certainly contain much of what you need to know. But, if not, it will give you the background to know where and how to look for what you do need, and to understand what you are doing when you apply a statistical method and reach a conclusion.

The book provides answers to most of the problems in the book. While this book is not meant to accompany a strictly computational course, calculations requiring a computer or at least a calculator are inevitably necessary. Therefore, this course requires the use of a TI-83/84/85/89, or any standard statistical package like Excel. Tables of things like the standard normal, t-distribution, etc., are not provided.

All experiments result in data. These data values are particular observations of underlying random variables. To analyze the data correctly the experimenter needs to be equipped with the tools that an understanding of probability and statistics provide. That is the purpose of this book. We will present basic, essential statistics, and the underlying probability theory to understand what the results mean.

The book begins with three foundational chapters on probability. The fundamental types of discrete and continuous random variables and their basic properties are introduced. Moment-generating functions are introduced at an early stage and used to calculate expected values and variances, and also to enable a proof of the Central Limit Theorem, a cornerstone result. Much of statistics is based on the Central Limit Theorem, and our view is that students should be exposed to a rigorous argument for why it is true and why the normal distribution plays such a central role. Distributions related to the normal distribution, like the χ^2, t, and F distributions, are presented for use in the statistical methods developed later in the book.

Chapter 3 is the prelude to the statistical topics included in the remainder of the text. This chapter includes the analysis of sample means and sample standard deviations as random variables.

The study of statistics begins in earnest with a discussion of confidence intervals in Chapter 4. Both one-sample and two-independent-samples confidence intervals are constructed as well as confidence intervals for paired data. Chapter 5 contains the topics at the core of statistics, particularly important for experimenters. We introduce tests of hypotheses for the major categories of experiments. Throughout the chapter, the dual relationship between confidence intervals and hypotheses tests is emphasized. The power of a test of hypotheses is discussed in some detail. Goodness-of-fit tests, contingency tables, tests for independence, and one-way analysis of variance is presented.

The book ends with a basic discussion of linear regression, an extremely useful tool in statistics. Calculus students have more than enough background to understand and appreciate how the results are derived. The probability theory introduced in earlier chapters is sufficient to analyze the coefficients derived in the regression.

The book has been used in the Introduction to Statistics & Probability course at our university for several years. It has been used in both two lectures per week and three lectures per week formats. Each semester typically involves at least two midterms (usually three) and a comprehensive final exam. In addition, class time includes time for group work as well as in-class quizzes. It makes it a busy semester to finish.

E.N. Barron and J.G. Del Greco
August 2020

Acknowledgments

We gratefully acknowledge Susanne Filler at Morgan & Claypool.

E.N. Barron and J.G. Del Greco
August 2020

CHAPTER 1

Probability

There are two kinds of events which occur: deterministic and stochastic (or random as a synonym). Deterministic phenomena are such that the same inputs always gives the exact same outputs. Newtonian physics deals with deterministic phenomena, but even real-world science is subject to random effects. Engineering design usually has a criteria which is to be met with 95% certainty because 100% certainty is impossible and too expensive. Think of designing an elevator in a high-rise building. Does the engineer know with certainty how many people will get on the elevator? What happens to the components of the elevator as they age? Do we know with certainty when they will collapse? These are examples of random events and we need a way to quantify them.

Statistics is based on probability which is a mathematical theory used to make sense out of random events and phenomena. In this chapter we will cover the basic concepts and techniques of probability we will use throughout this book.

1.1 THE BASICS

Every experiment whether it is designed by some experimenter or not, results in a possible **outcome**. If you deal five cards from a deck of cards, the hand you get is an outcome.

Definition 1.1   The **sample space** is the set S of all possible outcomes of an experiment.

S could be a finite set (like the number of all possible five card hands), a countably infinite set (like $\{0, 1, 2, 3 \ldots\}$ in a count of the number of users logging on to a computer system), or a continuum (like an interval $[a, b]$, like selecting a random number from a to b).

Definition 1.2   An **event** A is any subset of S, $A \subset S$.
The set S is also called the **sure event**. The empty set $\emptyset$ is called the **impossible event**.
The class of all possible events is denoted by $\mathcal{F} = \{A \mid A \subset S\}$.
If S is a finite set with N elements, we write $|S| = N$ and then the number of possible events is 2^N. (Why?)

Example 1.3

- If we roll a die the sample space is $S = \{1, 2, 3, 4, 5, 6\}$. Rolling an even number is the event $A = \{2, 4, 6\}$.

- If we want to count the number of customers coming to a bakery the sample space is $S = \{0, 1, 2, \dots\}$, and the event we get between 2 and 7 customers is $A = \{2, 3, 4, 5, 6, 7\}$.

- If we throw a dart randomly at a circular board of radius 2 feet, the sample space is the set of all possible positions of the dart $S = \{(x, y) \mid x^2 + y^2 \leq 4\}$. The event that the dart landed in the first quadrant is $A = \{(x, y) \mid x^2 + y^2 \leq 4, x \geq 0, y \geq 0.\}$.

Eventually we want to find the probability that an event will occur. We say that an event A **occurs** if any outcome in the set A actually occurs when the experiment is performed.

Combinations of events:

Let $A, B \in \mathcal{F}$ be any two events. From these events we may describe the following events:

(a) $A \cup B$ is the event A occurs, **or** B occurs, or they **both occur**.

(b) $A \cap B$, also written as AB, is the event A occurs **and** B occur, i.e., they both occur.

(c) $A^c = S - A$ is the event A does **not** occur. This is all the outcomes in S and not in A.

(d) $A \cap B^c$ is the event A occurs and B does not occur.

(e) $A \cap B = \emptyset$ means the two events **cannot occur together,** i.e., they are **mutually exclusive**. We also say that A and B are **disjoint**. Mutually exclusive events cannot occur at the same time.

(f) $A \cup A^c = S$ means that no matter what event A we pick, either A occurs or A^c occurs, and not both. A and A^c are mutually exclusive.

Many more such relations hold if we have three or more events. It is useful to recall the following set relationships.

- $A \cap (B \cup C) = (A \cap B) \cup (A \cap C)$ and $A \cup (B \cap C) = (A \cup B) \cap (A \cup C)$.

- $(A \cap B)^c = A^c \cup B^c$ and $(A \cup B)^c = A^c \cap B^c$. (DeMorgan's Rules)

These relations can be checked by using Venn diagrams.

Now we are ready to define what we mean by the probability of an event.

Definition 1.4   A **probability function** is a function $P : \mathcal{F} \to \mathbb{R}$ satisfying

- $\boxed{P(A) \geq 0, \ \forall A \in \mathcal{F}}$, probabilities cannot be negative,

- $\boxed{P(S) = 1}$, the probability of the sure event is 1,

- $\boxed{P(A \cup B) = P(A) + P(B)}$ for all events $A, B \in \mathcal{F}$ such that $A \cap B = \emptyset$.

This is called the **disjoint event sum rule**.

Whenever we write P we will always assume it is a probability function.

Remark 1.5 Immediately from the definition we can see that $P(\emptyset) = 0$. In fact, since we have the disjoint sum rule

$$S \cap \emptyset = \emptyset, P(S) = 1 = P(S \cup \emptyset) = P(S) + P(\emptyset) = 1 + P(\emptyset) \implies P(\emptyset) = 0.$$

Since $1 = P(S) = P(A \cup A^c) = P(A) + P(A^c)$ we also see that

$$\boxed{P(A^c) = 1 - P(A)}$$

for any event $A \in \mathcal{F}$.

It is also true that no matter what event $A \in \mathcal{F}$ we take $0 \le P(A) \le 1$. In fact, by definition $P(A^c) \ge 0$, and since $P(A^c) = 1 - P(A) \ge 0$, it must be that $0 \le P(A) \le 1$.

Remark 1.6 One of the most important and useful rules is the **Law of Total Probability**:

$$\boxed{P(A) = P(A \cap B) + P(A \cap B^c), \text{ for any events } A, B \in \mathcal{F}}. \tag{1.1}$$

To see why this is true, we use some basic set theory decomposing A,

$$A = A \cap S = A \cap (B \cup B^c) = (A \cap B) \cup (A \cap B^c)$$

and $A \cap B$ is disjoint from $A \cap B^c$. Therefore, by the disjoint event sum rule,

$$P(A) = P((A \cap B) \cup (A \cap B^c)) = P(A \cap B) + P(A \cap B^c).$$

A main use of this Law is that we may find the probability of an event A if we know what happens when $A \cap B$ occurs and when $A \cap B^c$ occurs. A useful form of this is $P(A \cap B^c) = P(A) - P(A \cap B)$.

The next theorem gives us the sum rule when the events are not mutually exclusive.

Theorem 1.7 **General Sum Rule.** If A, B are any two events, then

$$\boxed{P(A \cup B) = P(A) + P(B) - P(A \cap B).}$$

Proof. A union can always be written as a disjoint union. That is, $A \cup B = A \cup (A^c \cap B)$. Then by the disjoint sum rule $P(A \cup B) = P(A) + P(A^c \cap B)$. But, by the Law of Total Probability $P(A^c \cap B) = P(B) - P(A \cap B)$. Putting these together we have

$$P(A \cup B) = P(A) + P(A^c \cap B) = P(A) + P(B) - P(A \cap B).$$

$\square$

The next example gives one of the most important probability functions for finite sample spaces.

Example 1.8   When the sample space is finite, say $|S| = N$, and all **individual outcomes in S are equally likely**, we may define a function

$$P(A) = \frac{n(A)}{N}, \text{ where } n(A) = \text{number of outcomes in } A.$$

To see that this is a probability function we only have to verify the conditions of the definition.

- $0 \le P(A) \le 1$ since $0 \le n(A) \le N$ for any event $A \subset S$.

- $P(S) = \frac{n(S)}{N} = 1$.

- $P(A \cup B) = \frac{n(A \cup B)}{N} = \frac{n(A)+n(B)}{N} = P(A) + P(B)$, if $A \cap B = \emptyset$.

The requirement that individual outcomes be equally likely is essential. For example, suppose we roll two dice and sum the numbers on each die. We take the sample space $S = \{2, 3, 4, \ldots, 12\}$. If we use this sample space and we assume the outcomes are equally likely then we would get that $P(\text{roll a } 7) = 1/11$ which is clearly not correct. The problem is that with this sample space, the individual outcomes are not equally likely. If we want equally likely outcomes we need to change the sample space to account for the result on each die:

$$S = \{(1, 1), (1, 2), \ldots, (1, 6), (2, 1), (2, 2) \ldots, (2, 6), \ldots, (6, 1), (6, 2), \ldots, (6, 6)\}. \quad (1.2)$$

This sample space has 36 outcomes and the event of rolling a 7 is

$$A = \{(1, 6), (6, 1), (2, 5), (5, 2), (3, 4), (4, 3)\}.$$

Then $P(A) = P(\text{roll a } 7) = 6/36$ is the correct probability of rolling a 7.

Example 1.9   Whenever the sample space can easily be written it is often the best way to find probabilities. As an example, we roll two dice and we let D_1 denote the number on the first die and D_2 the number on the second. Suppose we want to find $P(D_1 > D_2)$. The easiest way to solve this is to write down the sample space as we did in (1.2) and then use the fact that each outcome is equally likely. We have

$$\{D_1 > D_2\} = \{(2, 1), (3, 2), (3, 1), (4, 3), (4, 2), (4, 1),$$
$$(5, 4), (5, 3), (5, 2), (5, 1), (6, 5), (6, 4), (6, 3), (6, 2), (6, 1)\}.$$

This event has 15 outcomes which means $P(D_1 > D_2) = \frac{15}{36}$.

1.2 CONDITIONAL PROBABILITY

It is important to take advantage of information about the occurrence of a given event in calculating the probability of a separate event. The way to do that is to use conditional probability.

Definition 1.10 The conditional probability of event A, given that event B has occurred is

$$\boxed{P(A|B) = \frac{P(A \cap B)}{P(B)} \text{ if } P(B) > 0.}$$

If $P(B) = 0$, B does not occur.

One of the justifications for this definition can be seen from the case when the sample space is finite (and equally likely individual outcomes). We have, if $|S| = N$,

$$\frac{n(A \cap B)}{n(B)} = \frac{\frac{n(A \cap B)}{N}}{\frac{n(B)}{N}} = \frac{P(A \cap B)}{P(B)} = P(A|B).$$

The left-most side of this string is the fraction of outcomes in $A \cap B$ from the event B. In other words, it is the probability of A **using the reduced sample space** B. That is, if the outcomes in S are equally likely, $P(A|B)$ is the proportion of outcomes in both A and B relative to the number of outcomes in B.

The introduction of conditional probability gives us the following which follows by rearranging the terms in the definition.

$$\boxed{\textbf{Multiplication Rule:} \quad P(A \cap B) = P(A|B)P(B) = P(B|A)P(A).}$$

Example 1.11 In a controlled experiment to see if a drug is effective 71 patients were given the drug (event D), while 75 were given a placebo (event D^c). A patient records a response (event R) or not (event R^c). The following table summarizes the results.

	Drug	Placebo	Subtotals	Probability
Response	26	13	39	0.267
No Response	45	62	107	0.733
Subtotals	71	75	146	
Probability	0.486	0.514		

This is called a **two-way or contingency table** .

The sample space consists of 146 outcomes of the type (Drug, Response), (Placebo, Response), (Drug, No Response), or (Placebo, No Response), assumed equally likely. The numbers in the table are recorded after the experiment is performed and we estimate the probability of each event. For instance,

$$P(D) = \frac{71}{146} = 0.486, \quad P(R) = \frac{39}{146},$$

and so on. For example, $P(R)$ is obtained from 39 of the equally likely chosen patients exhibit a response (whether to the drug or the placebo).

We can use the Law of Total Probability to also calculate these probabilities. If we want the chance that a randomly chosen patient will record a response we use the fact that $R = (R \cap D) \cup (R \cap D^c)$, so

$$P(R) = P(R \cap D) + P(R \cap D^c) = \frac{26}{146} + \frac{13}{146} = \frac{39}{146} = .267$$

and

$$P(D) = P(D \cap R) + P(D \cap R^c) = \frac{26}{146} + \frac{45}{146} = \frac{71}{146} = 0.486.$$

We may answer various questions using conditional probability.

- If we choose at random a patient and we observe that this patient exhibited a response, what is the chance this patient took the drug? This is

$$P(D|R) = \frac{26}{39} = \frac{P(D \cap R)}{P(R)} = \frac{\frac{26}{146}}{\frac{39}{146}}.$$

Using the reduced sample space R is how we got the first equality.

- If we choose a patient at random and we observe that this patient took the drug, what is the chance this patient exhibited a response? This is $P(R|D) = \frac{26}{71}$. Notice that $P(R|D) \neq P(D|R)$.

- Find $P(R^c|D) = \frac{45}{71}$. Observe that since $P(D) = P(R \cap D) + P(R^c \cap D)$, we have $P(R^c|D) = P(R^c \cap D)/P(D) = (P(D) - P(R \cap D)) = 1 - P(R|D)$.

Using the Law of Total Probability we get an important formula and tool for calculating probabilities of events.

Theorem 1.12 $\boxed{P(A) = P(A|B)P(B) + P(A|B^c)P(B^c).}$

Proof. The Law of Total Probability combined with the multiplication rule says

$$P(A) = P(A \cap B) + P(A \cap B^c) = P(A|B)P(B) + P(A|B^c)P(B^c),$$

which is the statement in the theorem. □

Frequently, problems arise in which we want to find the conditional probability of some event and we have yet another event we want to take into account. The next corollary tells us how to do that.

Corollary 1.13 Let A, B, C be three events. Then

$$P(A|B) = P(A|B \cap C)P(C|B) + P(A|B \cap C^c)P(C^c|B),$$

assuming each conditional probability is defined.

Proof. Simply write out each term and use the theorem.

$$\begin{aligned} P(A \cap B) &= P(A \cap B \cap C) + P(A \cap B \cap C^c) \\ &= \frac{P(A \cap B \cap C)}{P(B \cap C)}P(B \cap C) + \frac{P(A \cap B \cap C^c)}{P(B \cap C^c)}P(B \cap C^c) \\ &= P(A|B \cap C)P(B \cap C) + P(A|B \cap C^c)P(B \cap C^c). \end{aligned}$$

Divide both sides by $P(B) > 0$ to get

$$\begin{aligned} \frac{P(A \cap B)}{P(B)} = P(A|B) &= P(A|B \cap C)\frac{P(B \cap C)}{P(B)} + P(A|B \cap C^c)\frac{P(B \cap C^c)}{P(B)} \\ &= P(A|B \cap C)P(C|B) + P(A|B \cap C^c)P(C^c|B). \end{aligned}$$

□

Another very useful fact is that conditional probabilities are actually probabilities and therefore all rules for probabilities apply to conditional probabilities as long as the given information remains the same.

Corollary 1.14 Let B be an event with $P(B) > 0$. Then $Q(A) = P(A|B)$, $A \in \mathcal{F}$, is a probability function.

Proof. We have to verify $Q(\cdot)$ satisfies the axioms of Definition 1.4. Clearly, $Q(A) \geq 0$ for any event A and $Q(S) = P(S|B) = \frac{P(S \cap B)}{P(B)} = \frac{P(B)}{P(B)} = 1$. Finally, let $A_1 \cap A_2 = \emptyset$,

$$P(A_1 \cup A_2 | B) = \frac{P((A_1 \cup A_2) \cap B)}{P(B)} = \frac{P((A_1 \cap B) \cup (A_2 \cap B))}{P(B)}$$
$$= \frac{P(A_1 \cap B) + P(A_2 \cap B)}{P(B)} = P(A_1|B) + P(A_2|B).$$

This means the disjoint sum rule holds. $\square$

Conditional probability naturally leads us to what it means when information about B doesn't help with the probability of A. This is an important concept and will be very helpful throughout probability and statistics.

Definition 1.15   Two events A, B are said to be **independent**, if the knowledge that one of the events occurred does not affect the probability that the other event occurs. That is,

$$\boxed{P(A|B) = P(A) \text{ and } P(B|A) = P(A).}$$

Using the definition of conditional probability, an equivalent definition is $P(A \cap B) = P(A)P(B)$.

Example 1.16   1. Suppose an experiment has two possible outcomes a, b, so the sample space is $S = \{a, b\}$. Suppose $P(a) = p$ and $P(b) = 1 - p$. If we perform this experiment $n \geq 1$ times with identical conditions from experiment to experiment, then the events of individual experiments are independent. We may calculate

$$P(n\,a's \text{ in a row}) = p^n, \quad P(n\,a's \text{ and then } b) = p^n(1 - p).$$

In particular, the chance of getting five straight heads in five tosses of a fair coin is $(\frac{1}{2})^5 = 1/32$.
   2. The following two-way table contains data on place of residence and political leaning.

	Moderate	Conservative	Total
Urban	200	100	300
Rural	75	225	300
Total	275	325	600

Is one's political leaning independent of place of residence? To answer this question, let $U = \{$urban$\}$, $R = \{$rural$\}$, $M = \{$moderate$\}$, $C = \{$conservative$\}$. Then $P(U \cap M) = 200/600 = 1/3$, $P(U) = 300/600 = 1/2$, $P(M) = 275/600$. Since $P(U \cap M) \neq P(U) \times P(M)$, they are not independent.

When events are not independent we can frequently use the information about the occurrence of one of the events to find the probability of the other. That is the basis of conditional probability. The next concept allows us to calculate the probability of an event if the entire sample space is split (or partitioned) into pieces and decomposing the event we are interested in into the parts occurring in each piece. Here's the idea.

If we have events $B_1, \ldots, B_n$ such that $B_i \cap B_j = \emptyset$, for all i, j and $\cup_{i=1}^n B_i = S$, then the collection $\{B_i\}_{i=1}^n$ is called a **partition of** S. In this case, the Law of Total Probability says

$$P(A) = \sum_{i=1}^n P(A \cap B_i), \text{ and } P(A) = \sum_{i=1}^n P(A|B_i)P(B_i)$$

for any event $A \in \mathcal{F}$. We can calculate the probability of an event by using the pieces of A that intersect each B_i. It is always possible to partition S by taking any event B and the event B^c. Then for any other event A,

$$P(A) = P(A \cap B) + P(A \cap B^c) = P(A|B)P(B) + P(A|B^c)P(B^c).$$

Example 1.17 Suppose we draw the second card from the top of a well-shuffled deck. We want to know the probability that this card is an Ace.

This seems to depend on what the first card is. Let $B = \{1\text{st card is an Ace}\}$ and consider the partition $\{B, B^c\}$. We condition on what the first card is.

$$
\begin{aligned}
P(2\text{nd card is an Ace}) &= P(2\text{nd and 1st are Aces}) \\
&\quad + P(1\text{st is not an Ace and 2nd is an Ace}) \\
&= P(2\text{nd is Ace}|B)P(B) + P(2\text{nd is Ace}|B^c)P(B^c) \\
&= \frac{3}{51} \times \frac{4}{52} + \frac{4}{51} \times \frac{48}{52} = \frac{4}{52}.
\end{aligned}
$$

Amazingly, the chances the *second* card is an ace is the same as the chance the first card is an ace. This makes sense because if we don't know what the first card is, the second card should have the same chance as the first card. In fact, the chance the 27th card is an ace is also $4/52$ as long as we don't know any of the preceding 26 cards.

The next important theorem tells us how to find $P(B_k|A)$ if we know how to find $P(A|B_i)$ for each event B_i in the partition of S. It shows us how to find the probability that if A occurs, it was due to B_k.

Theorem 1.18 **Bayes' Rule.** Let $\{B_i\}_{i=1}^n$ be a partition of S. Then for each $k = 1, 2, \ldots, n$.

$$P(B_k|A) = \frac{P(B_k \cap A)}{P(A)} = \frac{P(A|B_k)P(B_k)}{P(A)} = \frac{P(A|B_k)P(B_k)}{P(A|B_1)P(B_1) + \cdots + P(A|B_n)P(B_n)}.$$

The proof is in the statement of the theorem using the definition of conditional probability and the Law of Total Probability.

Example 1.19   This example shows the use of both the Law of Total Probability and Bayes' Rule. Suppose there is a box with 10 coins, 9 of which are fair coins (probability of heads is 1/2), and 1 of which has heads on both sides. Suppose a coin is picked at random and it is tossed 5 times. Given that all 5 tosses result in heads, what is the probability the 6th toss will be a head?

Let $A = \{\text{toss 6 is a H}\}$, $B = \{\text{1st 5 tosses are H}\}$, and $C = \{\text{coin chosen is fair}\}$. The problem is we can't calculate $P(A)$ or $P(B)$ until we know what kind of coin we have. We need to condition on the type of coin. Here's what we know. $P(C) = 9/10$, and

$$P(A|C) = P(\text{toss 6 is a H}|\text{coin chosen is fair}) = \frac{1}{2}$$
$$P(A|C^c) = P(\text{toss 6 is a H}|\text{coin chosen is not fair}) = 1.$$

Now let's use Corrollary 1.13:

$$
\begin{aligned}
P(A|B) &= P(A|B \cap C)P(C|B) + P(A|B \cap C^c)P(C^c|B) \\
&= \frac{1}{2}\frac{P(B|C)P(C)}{P(B|C)P(C) + P(B|C^c)P(C^c)} + 1 \times (1 - P(C|B)) \\
&\qquad\qquad\qquad \text{using Bayes' Formula and } P(C^c|B) = 1 - P(C|B) \\
&= \frac{1}{2}\frac{(1/2)^5 \times 9/10}{(1/2)^5 \times 9/10 + 1 \times (1/10)} + 1 \times (1 - P(C|B)) \\
&= \frac{1}{2}\frac{9}{41} + \frac{32}{41} = \frac{73}{82} = 0.8902.
\end{aligned}
$$

Example 1.20   Tests for a medical condition are not foolproof. To see what this implies, suppose a test for a virus has sensitivity 0.95 and specificity 0.92. This means

$$\text{Sensitivity} = P(\text{test positive}|\text{have the disease}) = P(TP|D) = 0.95$$

and

$$\text{Specificity} = P(\text{test negative}|\text{do not have the disease}) = P(TP^c|D^c) = 0.92.$$

Suppose the prevalence of the disease is 5%, which means $P(D) = 0.05$. The question is if someone tests positive for the disease, what are the chances this person actually has the disease?

This is asking for $P(D|TP)$ but what we know is $P(TP|D)$ and $P(TP^c|D^c)$. This is a perfect use of Bayes' rule. We also use Corrollary 1.14:

$$
\begin{aligned}
P(D|TP) &= \frac{P(D \cap TP)}{P(TP)} = \frac{P(TP|D)P(D)}{P(TP)} \\
&= \frac{P(TP|D)P(D)}{P(TP|D)P(D) + P(TP|D^c)P(D^c)} \\
&= \frac{P(TP|D)P(D)}{P(TP|D)P(D) + (1 - P(TP^c|D^c))P(D^c)} \\
&= \frac{0.95 \times 0.05}{0.95 \times 0.05 + (1 - 0.92) \times 0.95} = 0.3846.
\end{aligned}
$$

This is amazing. Only 38% of people who test positive actually have the disease.

Example 1.21   Suppose there is a 1% chance of contracting a rare disease. Let D be the event you have the disease and TP the event you test positive for the disease. We know $P(TP|D) = 0.98$, and $P(TP^c|D^c) = 0.95$. As in the previous example, we first ask: given that you test positive, what is the probability that you really have the disease? We know how to work this out:

$$
P(D|TP) = \frac{0.98(0.01)}{0.98(0.01) + (1 - 0.95)(0.99)} = 0.165261.
$$

Now suppose there is an **independent** repetition of the test. Suppose the second test is also positive and now you want to know the probability that you really have the disease given the two positives.

To solve this let $TP_i, i = 1, 2$ denote the event you test positive on test $i = 1, 2$. These events are assumed **conditionally independent**.[1] Therefore, again by Bayes' formula we have

$$
\begin{aligned}
P(D|TP_1 \cap TP_2) &= \frac{P(TP_1 \cap TP_2|D)P(D)}{P(TP_1 \cap TP_2)} \\
&= \frac{P(TP_1 \cap TP_2|D)P(D)}{P(TP_1 \cap TP_2|D)P(D) + P(TP_1 \cap TP_2|D^c)P(D^c)} \\
&= \frac{P(TP_1|D)P(TP_2|D)P(D)}{P(TP_1|D)P(TP_2|D)P(D) + (1 - P(TP_1^c|D^c))(1 - P(TP_2^c|D))P(D^c)} = 0.795099.
\end{aligned}
$$

This says that the patient who tests positive twice now has an almost 80% chance of actually having the disease.

Example 1.22   **Simpson's Paradox.** Suppose a college has two majors, A and B. There are 2000 male applicants to the college with half applying to each major. There are 1100 female

[1]Conditional independence means independent conditioned on some event, i.e., $P(A \cap B|C) = P(A|C) \times P(B|C)$. In our case, conditional independence means $P(TP_1 \cap TP_2|D) = P(TP_1|D)P(TP_2|D)$.

applicants with 100 applying to A, and the rest to B. Major A admits 60% of applicants while major B admits 30%. This means that the percentage of men and women who apply to the college must be the same, right? Wrong.

In fact, we know that a total of 900 male applicants to the college were admitted giving 900/2000=0.45 or 45% of men admitted. For women the percentage is 360/1100=0.327 or 33%. Aggregating the men and women covers the fact that a larger percentage of women applied to the major which has a lower acceptance rate. This is an example of Simpson's paradox. Here's another example.

Two doctors have a record of success in two types of surgery, Low Risk and High Risk. Here's the table summarizing the results.

	Doctor A	Doctor B
Low Risk	93% (81/87)	87% (234/270)
High Risk	73% (192/263)	69% (55/80)
Total	78% (273/350)	83% (289/350)

The data show that conditioned on either low- or high-risk surgeries, Doctor A has a better success percentage. However, aggregating the high- and low-risk groups together produces the **opposite conclusion**. The explanation of this is arithmetic, namely, for numbers a, b, c, d, A, B, C, D, it is not true that $\frac{A}{B} > \frac{a}{b}$ and $\frac{C}{D} > \frac{c}{d}$ implies $\frac{A+C}{B+D} > \frac{a+c}{b+d}$. In the example, $81/87 > 234/270$ and $192/263 > 55/80$ but $(81 + 192)/(87 + 263) < (234 + 55)/(270 + 80)$.

1.3   APPENDIX: COUNTING

When we have a finite sample space $|S| = N$ with equally likely outcomes, we calculate $P(A) = \frac{n(A)}{N}$. It is sometimes a very difficult task to calculate both N and $n(A)$. In this appendix we give some basic counting principles to help with this.

Basic Counting Principle:
If there is a task with two steps and step one can be done in k different ways, and step two in j different ways, then the task can be completed in $k \times j$ different ways.

Permutations:
The number of ways to arrange k objects out of n distinct objects is

$$n(n-1)(n-2)\cdots(n-(k-1)) = \frac{n!}{(n-k)!} = P_{n,k}.$$

For instance, if we have 3 distinct objects $\{a, b, c\}$, there are $6 = 3 \cdot 2$ ways to pick 2 objects out of the 3, since there are 3 ways to pick the first object and then 2 ways to pick the second. They are $(a, b), (a, c), (b, a), (b, c), (c, a), (c, b)$.

Combinations:
The number of ways to choose k objects out of n when we don't care about the order of the objects is

$$C_{n,k} = \frac{n!}{k!(n-k)!} = \binom{n}{k}.$$

For example, in the paragraph on permutations, the choices (a, b) and (b, a) are different permutations but they are the same combination and so should not be counted separately. The way to get the number of combinations is to first figure out the number of permutations, namely $\frac{n!}{(n-k)!}$, and then get rid of the number of ways to arrange the selection of k objects, namely $k!$. In other words,

$$P_{n.k} = \binom{n}{k} k! \implies \binom{n}{k} = \frac{n!}{(n-k)!k!}.$$

Example 1.23   Poker Hands. We will calculate the probability of obtaining some of the common 5 card poker hands to illustrate the counting principles. A standard 52-card deck has 4 suits (Hearts, Clubs, Spades, Diamonds) with each suit consisting of 13 cards labeled $2, 3, 4, 5, 6, 7, 8, 9, 10, J, Q, K, A$. Five cards from the deck are chosen at random (without replacement). We now want to find the probabilities of various poker hands.

The sample space S is all possible 5-card hands where order of the cards does not matter. These are combinations of 5 cards from the 52, and there are $\binom{52}{5} = 2{,}598{,}960 = |S| = N$ possible hands, all of which are equally likely.

Probability of a Royal Flush, which is $A, K, Q, J, 10$ all the same suit. Let $A = \{\text{royal flush}\}$. How many royal flushes are there? It should be obvious there are exactly 4, one for each suit. Therefore, $P(A) = 4/\binom{52}{5} = 0.00000153908$, an extremely rare event.

Probability of a Full House, which is 3 of a kind and a pair. Let $A = \{\text{full house}\}$. To get a full house we break this down into steps.

(a) Pick a card for the 3 of a kind. There are 13 types one could choose.

(b) Choose 3 out of the 4 cards of the same type chosen in the first step. There are $\binom{4}{3}$ ways to do that.

(c) Choose another type distinct from the first type. There are 12 ways to do that.

(d) Choose 2 cards of the same type chosen in the previous step. There are $\binom{4}{2}$ ways to do that.

We conclude that the number of full house hands is $n(A) = 13 \times \binom{4}{3} \times 12 \times \binom{4}{2} = 3744$. Consequently $P(A) = 3744/\binom{52}{5} = 0.00144$.

Probability of 3 of a Kind. This is a hand of the form $aaabc$ where b, c are cards neither of which has the same face value as a. Let A be the event we get 3 of a kind. The number of hands in A is calculated using the multiplication rule with these steps:

(a) Choose a card type

(b) Choose 3 of that type

(c) Choose 2 more distinct types

(d) Choose 1 card of each type

The number of ways to do this is $\binom{13}{1} \times \binom{4}{3} \times \binom{12}{2} \times \binom{4}{1} \times \binom{4}{1} = 54{,}912$, and so $P(A) = 54{,}912/\binom{52}{5} = 0.0211$

Probability of a Pair. This is a hand like $aabcd$ where b, c, d, are distinct cards without the same face values. A is the event that we get a pair. To get one pair and make sure the other 3 cards don't match the pair is a bit tricky.

(a) Choose a card type

(b) Choose 2 of that type

(c) Choose 3 types from the remaining types

(d) Choose 1 card from each of these types

The number of ways to do that is $\binom{13}{1} \times \binom{4}{2} \times \binom{12}{3} \times \binom{4}{1}^3 = 1{,}098{,}240$, Therefore, $P(A) = 0.4225$. An exercise asks for the probability of getting 2 pairs.

1.4   PROBLEMS

1.1. Suppose $P(A) = p$, $P(B) = 0.3$, $P(A \cup B) = 0.6$. Find p so that $P(A \cap B) = 0$. Also, find p so that A and B are independent.

1.2. When $P(A) = 1/3$, $P(B) = 1/2$, $P(A \cup B) = 3/4$, what is (a) $P(A \cap B)$? and (b) what is $P(A^c \cup B)$?

1.3. Show $P(AB^c) = P(A) - P(AB)$ and P (exactly one of A or B occur) $= P(A) + P(B) - 2P(A \cap B)$.

1.4. 32% of Americans smoke cigarettes, 11% smoke cigars, 7% smoke both.

(a) What percent smoke neither cigars nor cigarettes?

(b) What percent smoke cigars but not cigarettes?

1.5. Let A, B, C be events. Write the expression for

(a) only A occurs

(b) both A and C occur but not B

(c) at least one occurs

(d) at least 2 occur

(e) all 3 occur

(f) none occur

(g) at most 2 occur

(h) at most 1 occurs

(i) exactly 2 occur

(j) at most 3 occur

1.6. Suppose $n(A)$ is the number of times A occurs if an experiment is performed N times. Set $F_N(A) = \frac{n(A)}{N}$. Show that F_N satisfies the definition to be a probability function. This leads to the frequency definition of the probability of an event $P(A) = \lim_{N \to \infty} F_N(A)$, i.e., the probability of an event is the long term fraction of time the event occurs.

1.7. Three events A, B, C cannot occur simultaneously. Further it is known that $P(A \cap B) = P(B \cap C) = P(A \cap C) = 1/3$. Can you determine $P(A)$? Hint: $A \subset (B \cap C)^c$.

1.8. (a) Give an example to illustrate $P(A) + P(B) = 1$ does not imply $A \cap B = \emptyset$. (b) Give an example to illustrate $P(A \cup B) = 1$ does not imply $A \cap B = \emptyset$. (c) Prove that $P(A) + P(B) + P(C) = 1$ if and only if $P(AB) = P(AC) = P(BC) = 0$.

1.9. A box contains 2 white balls and an unknown amount (finite) of non-white balls. Suppose 4 balls are chosen at random without replacement and suppose the probability of the sample containing both white balls is twice the probability of the sample containing no white balls. Find the total number of balls in the box.

1.10. Let C and D be two events for which one knows that $P(C) = 0.3$, $P(D) = 0.4$, $P(C \cap D) = 0.2$. What is $P(C^c \cap D)$?

1.11. An experiment has only two possible outcomes, only one of which may occur. The first has probability p to occur, the second probability p^2. What is p?

1.12. We repeatedly toss a coin. A head has probability p, and a tail probability $1 - p$ to occur, where $0 < p < 1$. What is the probability the first head occurs on the 5th toss? What is the probability it takes 5 tosses to get two heads?

1.13. Show that if $A \subset B$, then $P(A) \le P(B)$.

1.14. Analogously to the finite sample space case with equally likely outcomes we may define $P(A) = \text{area of } A / \text{area of } S$, where $S \subset \mathbb{R}^2$ is a fixed two-dimensional set (with equally likey outcomes) and $A \subset S$. Suppose that we have a dart board given by $S = \{x^2 + y^2 \le 9\}$ and A is the event that a randomly thrown dart lands in the ring with inner radius 1 and outer radius 2. Find $P(A)$.

1.15. Show that $P(A \cup B \cup C) = P(A) + P(B) + P(C) - P(A \cap B) - P(A \cap C) - P(C \cap B) + P(A \cap B \cap C)$.

1.16. Show that $P(A \cap B) \geq P(A) + P(B) - 1$ for all events $A, B \in \mathcal{F}$. Use this to find a lower bound on the probability both events occur if the probability of each event is 0.9.

1.17. A fair coin is flipped twice. We know that one of the tosses is a Head. Find the probability the other toss is a Head. (Hint: The answer is not 1/2.).

1.18. Find the probability of two pair in a 5-card poker hand.

1.19. Show that DeMorgan's Laws $(A \cup B)^c = A^c \cap B^c$ and $(A \cap B)^c = A^c \cup B^c$ hold and then find the probability neither A nor B occur and the probability either A does not occur or B does not but one of the two does occur. Your answer should express these in terms of $P(A), P(B)$, and $P(A \cap B)$.

1.20. Show that if A and B are independent events, then so are A and B^c as well as A^c and B^c.

1.21. If $P(A) = \dfrac{1}{3}$ and $P(B^c) = \dfrac{1}{4}$ is it possible that $A \cap B = \emptyset$? Explain.

1.22. Suppose we choose one of two coins C_1 or C_2 in which the probability of getting a head with C_1 is 1/3, and with C_2 is 2/3. If we choose a coin at random what is the probability we get a head when we flip it?

1.23. Suppose two cards are dealt one at a time from a well-shuffled standard deck of cards. Cards are ranked $2 < 3 < \cdots < 10 < J < Q < K < A$.

  (a) Find the probability the second card beats the first card. Hint: Look at $\sum_k P(C_2 > C_1 | C_1 = k) P(C_1 = k)$.
  (b) Find the probability the first card beats the second and the probability the two cards match.

1.24. A basketball team wins 60% of its games when it leads at the end of the first quarter, and loses 90% of its games when the opposing team leads. If the team leads at the end of the first quarter about 30% of the time, what fraction of the games does it win?

1.25. Suppose there is a box with 10 coins, 8 of which are fair coins (probability of heads is 1/2), and 2 of which have heads on both sides. Suppose a coin is picked at random and it is tossed 5 times. Given that we got 5 straight heads, what are the chances the coin has heads on both sides?

1.26. Is independence for three events A, B, C the same as: A, B are independent; B, C are independent; and A, C are independent? Consider the example: Perform two independent tosses of a coin. Let A =heads on toss 1, B =heads on toss 2, and C =the two tosses are equal.

(a) Find $P(A)$, $P(B)$, $P(C)$, $P(C|A)$, $P(B|A)$, and $P(C|B)$. What do you conclude?

(b) Find $P(A \cap B \cap C)$ and $P(A \cap B \cap C^c)$. What do you conclude?

1.27. First show that $P(A \cup B) = P(A) + P(A^c \cap B)$ and then calculate.

(a) $P(A \cup B)$ if it is given that $P(A) = 1/3$ and $P(B|A^c) = 1/4$.

(b) $P(B)$ if it is given that $P(A \cup B) = 2/3$, $P(A^c|B^c) = 1/2$.

1.28. The events A, B, and C satisfy: $P(A|B \cap C) = 1/4$, $P(B|C) = 1/3$, and $P(C) = 1/2$. Calculate $P(A^c \cap B \cap C)$.

1.29. Two independent events A and B are given, and $P(B|A \cup B) = 2/3$, $P(A|B) = 1/2$. What is $P(B)$?

1.30. You roll a die and a friend tosses a coin. If you roll a 6, you win. If you don't roll a 6 and your friend tosses a H, you lose. If you don't roll a 6, and your friend does not toss a H, the game repeats. Find the probability you Win.

1.31. You are diagnosed with an uncommon disease. You know that there only is a 4% chance of having the disease. Let D =you have the disease, and T =the test says you have it. It is known that the test is imperfect: $P(T|D) = 0.9$ and $P(T^c|D^c) = 0.85$.

(a) Given that you test positive, what is the probability that you really have the disease?

(b) You obtain a second and third opinion: **two more (conditionally) independent repetitions of the test**. You test positive again on both tests. Assuming conditional independence, what is the probability that you really have the disease?

1.32. Two dice are rolled. What is the probability that at least one is a six? If the two faces are different, what is the probability that at least one is a six?

1.33. 15% of a group are heavy smokers, 30% are light smokers, 55% are nonsmokers. In a 5-year study it was determined that the death rates of heavy and light smokers were 5 and 3 times that of nonsmokers, respectively. What is the probability a randomly selected person was a nonsmoker, given that he died?

1.34. A, B, and C are mutually independent and $P(A) = 0.5$, $P(B) = 0.8$, $P(C) = 0.9$. Find the probabilities (i) all three occur, (ii) exactly 2 of the 3 occur, or (iii) none occurs.

1.35. A box has 8 red and 7 blue balls. A second box has an unknown number of red and 9 blue balls. If we draw a ball from each box at random we know the probability of getting 2 balls of the same color is 151/300. How many red balls are in the second box?

1.36. Show that:

(a) $P(A|A \cup B) \geq P(A|B)$. Hint: $A = (A \cap B) \cup (A \cap B^c)$ and $A \cup B = B \cup (A \cap B^c)$.

(b) If $P(A|B) = 1$ then $P(B^c|A^c) = 1$.

(c) $P(A|B) \geq P(A) \implies P(B|A) \geq P(B)$.

1.37. Coin 1 has H with probability 0.4; Coin 2 has H with probability 0.7. One of these coins is chosen at random and flipped 10 times. Find

(a) P (coin lands H on exactly 7 of the 10 flips) and

(b) given the first of these 10 flips is H, find the conditional probability that exactly 7 of the 10 flips is H.

1.38. Show the extended version of the Law of Total Conditional Probability

$$P(A|B) = \sum_i P(A|E_i \cap B)P(E_i|B), \ \ S = \cup_i E_i, \ E_i \cap E_j = \emptyset \ i \neq j.$$

1.39. There are two universities. The breakdown of males and females majoring in Math at each university is given in the tables.

Univ 1	Math Major	Other
Males	200	800
Females	150	850

Univ 2	Math Major	Other
Males	30	70
Females	1000	3000

Show that this is an example of Simpson's paradox.

1.40. The table gives the result of a drug trial:

	M Recover	M Die	F Recover	F Die	O Recover	O Die
Drug	15	40	90	50	105	90
No Drug	20	40	20	10	40	50

Here M = male, F = female, and O = overall. Show that this is an example of Simpson's paradox.

CHAPTER 2

Random Variables

In this chapter we study the main properties of functions whose domain is an outcome of an experiment with random outcomes, i.e., a sample space. Such functions are called random variables.

2.1 DISTRIBUTIONS

The distribution of a random variable is a specification of the probability that the random variable takes on any set of values. What is a random variable? It is just a function defined on the sample space S of an experiment.

Definition 2.1 A random variable (rv) is a function $X : S \to \mathbb{R}$ such that $E = \{s \in S \mid X(s) \le a\} \in \mathcal{F}$, the set of all possible events, for every real number a. In other words, we want every set of the type $\{s \in S \mid X(s) \le a\}$ to be an event.

As a function, a random variable has a range $R(X) = \{y \in \mathbb{R} \mid X(s) = y, \text{ for some } s \in S\}$. If $R(X)$ is a finite or countable set, we say X is **discrete**. If $R(X)$ contains an interval, then we say X is not discrete, but either **continuous**, or mixed.

Definition 2.2 If X is a **discrete** random variable with range $R(X) = \{x_1, x_2, \ldots, \}$, the **probability mass function** (pmf) of X is $p(x_i) = P(X = x_i), i = 1, 2, \ldots$. We write[1] $\{X = x_i\}$ for the event $\{X = x_i\} = \{s \in S \mid X(s) = x_i\}$.

Remark 2.3 Any function $p(x_i)$ which satisfies (i) $0 \le p(x_i) \le 1$ for all i, and (ii) $\sum_i p(x_i) = 1$ is called a pmf. The pmf of a rv is also called its **distribution**.

The next two particular discrete rvs are fundamental.

Distribution 2.4 A rv X which takes on only two values, a, b, with $P(X = a) = p, P(X = b) = 1 - p$, is said to have a **Bernoulli**(a, b, p) **distribution**, or be a Bernoulli(a, b, p) rv, and we write $X \sim \text{Bernoulli}(a, b, p)$. In particular, if we have an experiment with two outcomes, success or failure, we may set $a = 1, b = 0$ to represent these, and p is the probability of success.

---

[1]In general, write $\{X \le a\} = \{s \in S \mid X(s) \le a\}$ and similarly for $\{X > a\}$.

An experiment like this is called a Bernoulli(p) trial. The pmf of a Bernoulli(a, b, p) rv is

$$P(X = x) = p(x) = \begin{cases} p, & \text{if } x = a; \\ 1 - p, & \text{if } x = b. \end{cases}$$

A rv X which **counts the number of successes** in an independent set of n Bernoulli trials is called a **Binomial**(n, p) rv, written $X \sim \text{Binom}(n, p)$. The range of X is $R(X) = \{0, 1, 2, \ldots, n\}$. The pmf of X is

$$P(X = x) = p(x) = \binom{n}{x} p^x (1 - p)^{n-x}, \quad x = 0, 1, 2 \ldots, n.$$

Remark 2.5   Here's where this comes from. If we have a particular sequence of n Bernoulli trials with x successes, say $10011101 \ldots 1$, then x 1's must be in this sequence and n-x 0's must also be in there. By independence of the trials, the probability of any **particular** sequence of x 1's and $n - x$ 0's is $p^x(1 - p)^{n-x}$. How many sequences with x 1's out of n are there? That number is $\binom{n}{x} = \dfrac{n!}{x!(n - x)!}$.

It should be clear that a Binomial(n, p) rv X is a sum of (independent) n Bernoulli(p) rvs, $X = X_1 + X_2 + \cdot + X_n$. Independent rvs will be discussed later.

Example 2.6   A bet on red for a standard roulette wheel has $\frac{18}{38}$ chances of winning. Suppose a gambler will bet \$5 on red each time for 100 plays. Let X be the total amount won or lost as a result of these 100 plays. X will be a discrete random variable with range $R(X) = \{0, \pm 5, \pm 10, \ldots, \pm 500\}$. In fact, if M denotes the number of games won (which is also a random variable with values from 0 to 100), then our net amount won or lost is $X = 10M - 500$. The random variable M is an example of a Binomial(100, 18/38) rv.

The chance you win exactly 50 games is $P(M = 50) = \frac{18}{38}^{50} \frac{20}{38}^{50} \binom{100}{50} = 0.0693$, so the chance you break even is $P(X = 0)$ is also 0.0693.

Now we define continuous random variables.

Definition 2.7   A random variable X is continuous if there is a function $f : \mathbb{R} \to \mathbb{R}$ associated with X such that $f(x) \geq 0$, for all x, and $\int_{-\infty}^{\infty} f(x)\, dx = 1$. The function f is called a **probability density function (pdf) of** X. It is important to note that a pdf does not have to satisfy $f(x) \leq 1$ in general.

Remark 2.8   Later it will turn out to be very useful to also use the notation for a pdf $f(x) = P(X = x)$ but we have to be careful with this because, as we will see, the probability a continuous

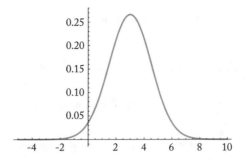

Figure 2.1: Normal Distribution with $\mu = 3, \sigma = 1.5$.

rv is any particular value is 0. This notation is purely to simplify statements and is intuitive as long as one keeps this in mind.

The next example defines the most important pdf in statistics.

Distribution 2.9 A rv X is said to have a **Normal distribution** with parameters $(\mu, \sigma), \sigma > 0$ if

$$f(x) = \frac{1}{\sigma\sqrt{2\pi}} \int_{-\infty}^{\infty} e^{-\frac{1}{2}\left(\frac{x-\mu}{\sigma}\right)^2} \, dx.$$

We write $X \sim N(\mu, \sigma)$ where, in general $\sim$ is to be read as **is distributed as**. If $\mu = 0, \sigma = 1$, the rv $Z \sim N(0, 1)$ is said to be a **standard normal** rv.

Figure 2.1 is a graph of the normal pdf with $\mu = 3, \sigma = 1.5$.

Remark 2.10 The line of symmetry of a $N(\mu, \sigma)$ is always at $x = \mu$, and it provides the point of maximum of the pdf. One can check this using the second derivative test and $f'(\mu) = 0, f''(\mu) < 0$. It is also a calculus exercise to check that $x = \mu + \sigma$ and $x = \mu - \sigma$ both provide points of inflection (where concavity changes) of the pdf.

Remark 2.11 It is not an easy task to check that $f(x)$ really is a density function. It is obviously always nonnegative but why is the integral equal to one? That fact uses the following formula which is verified in calculus,

$$\int_{-\infty}^{\infty} e^{-x^2} \, dx = \sqrt{\pi}.$$

Using this formula and a simple change of variables, one can verify that f is indeed a pdf.

How do we use pdfs to compute probabilities? Let's start with finding certain types of probabilities.

Definition 2.12   The cumulative distribution function (cdf) of a random variable X is $F_X(x) = P(X \leq x)$.

$$F_X(x) = \begin{cases} \sum_{x_i \leq x} p(x_i), & \text{if } X \text{ is } \textbf{discrete} \text{ with pmf } p, \\ \int_{-\infty}^{x} f(y) \, dy, & \text{if } X \text{ is } \textbf{continuous} \text{ with pdf } f \ . \end{cases}$$

Every cdf has the properties

(a) $\lim_{x \to -\infty} F_X(x) = 0$, and $\lim_{x \to +\infty} F_X(x) = 1$.

(b) $x < y \implies F_X(x) \leq F_X(y)$.(nondecreasing).

(c) $\lim_{y \to x+0} F_X(y) = F_X(x)$ for all $x \in \mathbb{R}$. This says a cdf is continuous at every point **from the right**.

Using the cdf F_X we have for $a < b$, $\boxed{P(a < X \leq b) = F_X(b) - F_X(a).}$
If X is continuous $P(a < X \leq b) = F_X(b) - F_X(a) = \int_a^b f(x) \, dx$.

Proof. $P(a < X \leq b) = P(\{X \leq b\} \cap \{X \leq a\}^c) = P(X \leq b) - P(X \leq a) = F_X(b) - F_X(a)$. We have used $P(B \cap A^c) = P(B) - P(A \cap B)$, $A = \{X \leq a\} \subset B = \{X \leq b\}$.   ☐

If X is continuous with density $f(x)$, $P(a < X \leq b) = \int_a^b f(x) \, dx$ represents the area under the density curve between a and b.

If X is discrete, X can be a single point with positive probability, $P(X = x) > 0$. If X is continuous $P(X = x) = 0$ for any x since

$$P(X = x) \leq P(x - \varepsilon < X \leq x) = F_X(x) - F_X(x - \varepsilon) = \int_{x-\varepsilon}^{x} f(y) \, dy \to 0 \text{ as } \varepsilon \to 0.$$

Therefore, for a continuous rv, $P(a < X \leq b) = P(a \leq X \leq b) = P(a \leq X < b)$ are all the same. For a discrete rv

$$P(a < X \leq b) = P(a < X < b) + P(X = b).$$

We have to take endpoints into account.

Remark 2.13   If X is continuous with pdf f and cdf F_X, we know that $F_X(x) = \int_{-\infty}^{x} f(y) \, dy$. Therefore, we can find the pdf if we know the cdf using the fundamental theorem of calculus, $F_X'(x) = f(x)$.

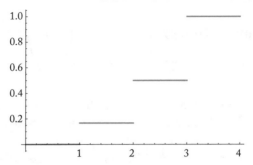

Figure 2.2: CDF Discrete $F_X(x)$.

Example 2.14 Suppose X is a random variable with values $1, 2, 3$ with probabilities $1/6, 1/3, 1/2$, respectively. In Figure 2.2 the jumps are at $x = 1, 2, 3$. The size of the jump is $P(X = x), x = 1, 2, 3$ and at each jump the left endpoint is not included while the right endpoint is included because the cdf is continuous from the right. Then we may calculate $P(X < 2) = P(X = 1) = 1/6$ but $P(X \le 2) = P(X = 2) + P(X = 1) = 1/2$.

2.2 IMPORTANT DISCRETE DISTRIBUTIONS

We begin with the pmfs of some of the most important discrete rvs we will use in this book.

Distribution 2.15 **Binomial**(n, p). $P(X = x) = \binom{n}{x} p^x (1 - p)^{n-x}, x = 0, 1, \ldots, n$. If $n = 1$ this is Bernoulli(p). A Binomial rv counts the number of successes in n independent Bernoulli trials.[2]

Distribution 2.16 **Discrete Uniform.** $P(X = x) = \frac{1}{n}, x = 1, 2, \ldots, n., F_X(x) = \frac{x}{n}, 1 \le x \le n$. A discrete uniform rv picks one of n points at random.

Distribution 2.17 **Poisson**(λ). $P(X = x) = e^{-\lambda} \frac{\lambda^x}{x!}, x = 0, 1, 2 \ldots$. The parameter $\lambda > 0$ is given. A Poisson(λ) rv counts the number of events that occur at the rate λ.[3]

Distribution 2.18 **Geometric**(p). $P(X = x) = (1 - p)^{x-1} p, x = 1, 2, \ldots$. X is the number of independent Bernoulli trials until we get the first success.[4]

---

[2]A Binomial pdf can be calculated on a TI-83/4/5 using binompdf$(n, p, k) = P(X = k)$ and binomcdf$(n, p, k) = P(X \le k)$.

[3]A Poisson pdf can be calculated using poissonpdf$(\lambda, k) = P(X = k)$ and poissoncdf$(\lambda, k) = P(X \le k)$.

[4]A Geometric pdf can be calculated using geometpdf$(p, k) = P(X = k)$ and geometcdf$(p, k) = P(X \le k)$.

Distribution 2.19   **NegBinomial**(r, p). $P(X = x) = \binom{x-1}{r-1} p^r (1-p)^{x-r}$, $x = r, r+1$,
$r+2, \ldots x$ represents the number of Bernoulli trials until we get r successes.

Distribution 2.20   **Hypergeometric**(N, n, k). Consider a population of total size N consisting of two types of objects, say Red and Black. Suppose there are k Red objects in the population and $N - k$ Black objects. We choose n objects from the population at random **without replacement.** X represents the number of Red objects we obtain in this process. We have

$$P(X = x) = \frac{\text{choose } x \text{ out of } k \text{ Reds} \times \text{choose } n - x \text{ out of } N - k \text{ Black}}{\text{choose } n \text{ out of } N} = \frac{\binom{k}{x} \times \binom{N-k}{n-x}}{\binom{N}{n}}.$$

Distribution 2.21   **Multinomial** $(n, p_1, \ldots, p_k)$. Suppose there is an experiment with n independent trials with k possible outcomes on each trial, labeled $A_1, A_2, \ldots, A_k$, with the probability of outcome i given by $p_i = P(A_i)$. Let X_i be a count of the number of occurrences of A_i. Then

$$P(X_1 = x_1, X_2 = x_2, \ldots, X_k = x_k) = \binom{n}{x_1, x_2, \ldots, x_k} p_1^{x_1} p_2^{x_2} \cdots p_k^{x_k},$$

where $x_1 + x_2 + \cdots + x_k = n$, $p_1 + p_2 + \cdots p_k = 1$. The multinomial coefficient is given by

$$\binom{n}{x_1, x_2, \ldots, x_k} = \frac{n!}{x_1! x_2! \cdots x_k!}.$$

It comes from choosing x_1 out of n, then x_2 out of $n - x_1$, then x_3 out of $n - x_1 - x_2$, etc.,

$$\binom{n}{x_1} \times \binom{n - x_1}{x_2} \times \cdots \times \binom{n - x_1 - x_2 - \cdots - x_{k-1}}{x_k} = \binom{n}{x_1, x_2, \ldots, x_k}.$$

This generalizes the binomial distribution to the case when there is more than just a success or failure on each trial.

The cdf $F_X(x)$ of each of these can be written down once we have the pmf. For example, for a Binom(n, p) rv, we have $F_X(x) = \sum_{k=0}^{x} \binom{n}{k} p^k (1-p)^{n-k} = \text{binomcdf}(n, p, x)$.

Example 2.22   For the Negative Binomial, X is the number of trials until we get r successes. We must have at least r trials to get r successes and we get r successes with probability p^r and $x - r$ failures with probability $(1-p)^{x-r}$. Since we stop counting when we get the rth success, the last trial must be a success. Therefore, in the preceding $x - 1$ trials we spread $r - 1$ successes and there are $\binom{x-1}{r-1}$ ways to do that. That's why $P(X = x) = \binom{x-1}{r-1} p^r (1 - p)^{x-r}, x \geq r$. Here is an example where the Negative Binomial arises.

Best of seven series. The baseball and NBA finals determines a winner by the two teams playing up to seven games with the first team to win four games the champ. Suppose team A wins with probability p each game and loses to team B with probability $1 - p$.

(a) If $p = 0.52$, what is the probability A wins the series? For A to win the series, A can win 4 straight, or in 5 games, or in 6 games, or in 7 games. This is negative binomial with $r = 4, 5, 6, 7$ so if X is the number of games to 4 successes for A,

$$P(A \text{ wins series}) = P(X = 4) + P(X = 5) + P(X = 6) + P(X = 7)$$

$$= \binom{3}{3}.52^4(.48)^0 + \binom{4}{3}.52^4(.48)^1 + \binom{5}{3}.52^4(.48)^2 + \binom{6}{3}.52^4(.48)^3$$

$$= 0.54368.$$

If $p = 0.55$, the probability A wins the series goes up to 0.60828 and if $p = 0.6$ the probability A wins is 0.7102.

(b) If $p = 0.52$ and A wins the first game, what is the probability A wins the series?

This is asking for $P(A \text{ wins series}|A \text{ wins game 1})$. Let X_1 be the number of games (out of the remaining 6) until A wins 3. Then

$$P(A \text{ wins 7 game series}|A \text{ wins game 1}) = \frac{P(A \text{ wins 7 game series} \cap A \text{ wins game 1})}{P(A \text{ wins game 1})}$$

$$= \frac{P(A \text{ wins 6 game series} \cap A \text{ wins game 1})}{P(A \text{ wins game 1})}$$

$$= \frac{pP(X_1 = 3) + pP(X_1 = 4) + pP(X_1 = 5) + pP(X_1 = 6)}{p}$$

$$= P(X_1 = 3) + P(X_1 = 4) + P(X_1 = 5) + P(X_1 = 6)$$

$$= \binom{2}{2}.52^3(.48)^0 + \binom{3}{2}.52^3(.48)^1 + \binom{4}{2}.52^3(.48)^2 + \binom{5}{2}.52^3(.48)^3 = \boxed{0.6929.}$$

An easy way to get this without conditional probability is to realize that once game one is over, A has to be the first to 3 wins in at most 6 trials.

Example 2.23   Multinomial distributions arise whenever one of two or more outcomes can occur. Here's a polling example. Suppose 25 registered voters are chosen at random from a population in which we know that 55% are Democrats, 40% are Republicans, and 5% are Independents. In our sample of 25, what are the chances we get 10 Democrats, 10 Republicans, and 5 Independents?

This is multinomial with $p_1 = .55, p_2 = .4, p_3 = 0.05$. Then

$$P(D = 10, R = 10, I = 5) = \binom{25}{10, 10, 5} .55^{10} .4^{10} .05^5 = 0.000814$$

which is really small because we are asking for exactly the numbers $(10, 10, 5)$. It is much more tedious to calculate but we can also find things like

$$P(D \leq 15, R \leq 12, I \leq 20) = 0.6038.$$

Notice that in the cumulative distribution we don't require that $15 + 12 + 20$ be the number of trials.

Example 2.24   Consider a random variable X with pdf $f_X(x) = \begin{cases} 60x^2(1-x)^3, & 0 \leq x \leq 1, \\ 0, & \text{otherwise.} \end{cases}$
Suppose 20 independent samples are drawn from X. An outcome is the sample value falling into range $[0, \frac{1}{5}]$ when $i = 1$ or $(\frac{i-1}{5}, \frac{i}{5}]$, $i = 2, 3, 4, 5$. What is the probability that 3 observations fall into the first range, 9 fall into the second range, 4 fall into the third and fourth ranges, and that there are no observations that fall into the fifth range? To answer this question, let p_i denote the probability of a sample value falling into range i. These probabilities are computed directly from the pdf. For example,

$$p_1 = \int_0^{0.2} 60x^2(1-x)^3 \, dx = 0.098.$$

Complete results are displayed in the table.

Range	$[0, 0.2]$	$(0.2, 0.4]$	$(0.4, 0.6]$	$(0.6, 0.8]$	$(0.8, 1.0]$
Probability	0.098	0.356	0.365	0.162	0.019

If X_i is the number of samples that fall into range i, we have

$$P(X_1 = 3, X_2 = 9, X_3 = 4, X_4 = 4, X_5 = 0)$$
$$= \binom{20}{3, 9, 4, 4, 0}(0.098)^3 (0.356)^9 (0.365)^4 (0.162)^4 (0.019)^0 = 0.00205.$$

Example 2.25   Suppose we have 10 patients, 7 of whom have a genetic marker for lung cancer and 3 of whom do not. We will choose 6 at random (without replacing them as we make our selection). What are the chances we get exactly 4 patients with the marker and 2 without?

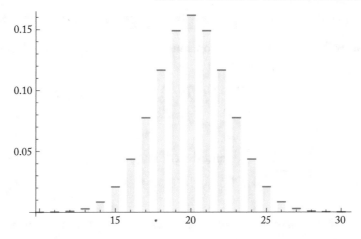

Figure 2.3: $P(X = k)$: Hypergeometric looks like normal if trials large enough.

This is Hypergeometric(10,6,7). View this as drawing 6 people at random from a group of 10 without replacement, with the probability of success (=genetic marker) changing from draw to draw. The trials are not Bernoulli. We have with X =number with genetic marker,

$$P(X = 4) = \frac{\binom{7}{4}\binom{3}{2}}{\binom{10}{6}} = \frac{1}{2}.$$

If we incorrectly assumed this was $X \sim \text{Binom}(6, 0.7)$ we would get $P(X = 4) = \text{binompdf}(6, 0.7, 4) = 0.324$.

As a further example, suppose we have a group of 100 with 40 patients possessing the genetic marker (=success). We draw 50 patients at random without replacement and ask for the $P(X = k), k = 10, 11, \ldots, 50$. Figure 2.3 shows the hypergeometric distribution.

The fact that the figure for the hypergeometric distribution looks like a normal curve is not a coincidence, as we will see, when the population is large.

2.3 IMPORTANT CONTINUOUS DISTRIBUTIONS

In this section we will describe the main continuous random variables used in probability and statistics.

Distribution 2.26   Uniform(a, b). $X \sim \text{Unif}(a, b)$ models choosing a random number from a to b. The pdf is

$$f(x) = \begin{cases} \dfrac{1}{b-a}, & \text{if } a < x < b; \\ 0, & \text{otherwise.} \end{cases} \quad \text{and the cdf is } F_X(x) = \begin{cases} 0, & \text{if } x < a; \\ \dfrac{x-a}{b-a}, & \text{if } a \leq x < b; \\ 1, & \text{if } x \geq b. \end{cases}$$

Next is the normal distribution which we have already discussed but we record it here again for convenience.

Distribution 2.27   Normal(μ, σ). $X \sim N(\mu, \sigma)$ has density

$$f(x) = \frac{1}{\sigma \sqrt{2\pi}} e^{-\frac{1}{2}\left(\frac{x-\mu}{\sigma}\right)^2}, \quad -\infty < x < \infty.$$

It is not possible to get an explicit expression for the cdf so we simply write

$$F_X(x) = N(x; \mu, \sigma) = \frac{1}{\sigma \sqrt{2\pi}} \int_{-\infty}^{x} e^{-\frac{1}{2}\left(\frac{y-\mu}{\sigma}\right)^2} dy.$$

We shall also write $P(a < X < b) = \text{normalcdf}(a, b, \mu, \sigma)$. [5]

Distribution 2.28   Exponential(λ). $X \sim Exp(\lambda), \lambda > 0$, has pdf $f(x) = \begin{cases} \lambda e^{-\lambda x}, & x \geq 0, \\ 0, & x < 0. \end{cases}$

The cdf is

$$F_X(x) = \int_0^x \lambda e^{-\lambda y} dy = \begin{cases} 1 - e^{-\lambda x}, & \text{if } x \geq 0; \\ 0, & \text{if } x < 0. \end{cases}$$

An exponential random variable represents processes that do not remember. For example, if X represents the time between arrivals of customers to a store, a reasonable model is Exponential(λ) where λ represents the average rate at which customers arrive.

At the end of this chapter we will introduce the remaining important distributions for statistics including the χ^2, t, and F-distributions. They are built on combinations of rvs. Now we look at an important transformation of a rv in the next example.

Example 2.29   Change of scale and shift. If we have a random variable X which has pdf f and cdf F_X we may calculate the pdf and cdf of the random variable $Y = \alpha X + \beta$, where $\alpha \neq 0, \beta$

---

[5]$\text{normalcdf}(a, b, \mu, \sigma)$ is a command from a TI-8x calculator which gives the area under the normal density with parameters μ, σ, from a to b.

are constants. To do so, start with the cdf:

$$F_Y(y) = P(Y \le y) = \begin{cases} P\left(X \le \dfrac{y-\beta}{\alpha}\right), & \alpha > 0, \\[2mm] P\left(X \ge \dfrac{y-\beta}{\alpha}\right), & \alpha < 0. \end{cases} = \begin{cases} F_X\left(\dfrac{y-\beta}{\alpha}\right), & \alpha > 0, \\[2mm] 1 - F_X\left(\dfrac{y-\beta}{\alpha}\right), & \alpha < 0. \end{cases}$$

Then we find the pdf by taking the derivative:

$$f_Y(y) = \frac{d}{dy} F_Y(y) = \begin{cases} f_X\left(\dfrac{y-\beta}{\alpha}\right)\dfrac{1}{\alpha}, & \alpha > 0, \\[3mm] -f_X\left(\dfrac{y-\beta}{\alpha}\right)\dfrac{1}{\alpha}, & \alpha < 0. \end{cases}$$

In particular, if $X \sim N(\mu, \sigma)$, we have the pdf for $Y = \alpha X + \beta$, (assuming $\alpha > 0$),

$$f_Y(y) = \frac{1}{\alpha\sigma\sqrt{2\pi}} e^{-\frac{1}{2}\left(\frac{y-\alpha\mu-\beta}{\alpha\sigma}\right)^2}$$

which we recognize as the pdf of a $N(\alpha\mu + \beta, \alpha\sigma)$ random variable. Thus, $Y = \alpha X + \beta \sim N(\alpha\mu + \beta, \alpha\sigma)$.

If we take $\alpha = \frac{1}{\sigma}, \beta = -\frac{\mu}{\sigma}, Y = \frac{1}{\sigma}X - \frac{\mu}{\sigma}$, then $Y \sim N(0, 1)$. We have shown that given any $X \sim N(\mu, \sigma)$, if we set

$$\boxed{Y = \frac{X - \mu}{\sigma} \implies Y \sim N(0, 1).}$$

The rv $X \sim N(\mu, \sigma)$ has been converted to the standard normal rv $Y \sim N(0, 1)$. Starting with $X \sim N(\mu, \sigma)$ and converting it to $Y \sim N(0, 1)$ is called **standardizing** X.

The reason that a normal distribution is so important is contained in the following special case of the Central Limit Theorem.

Theorem 2.30   Central Limit Theorem for Binomial. Let $S_n \sim \text{Binom}(n, p)$. Then

$$\lim_{n\to\infty} P\left(\frac{S_n - np}{\sqrt{np(1-p)}} \le x\right) = \frac{1}{\sqrt{2\pi}} \int_{-\infty}^{x} e^{-z^2/2}\, dz = \text{normcdf}(-\infty, x, 0, 1), \quad \text{for all } x \in \mathbb{R}.$$

In short, $\dfrac{S_n - np}{\sqrt{np(1-p)}} \approx N(0, 1)$ for large n. Alternatively, $S_n \approx N(np, \sqrt{np(1-p)}\,)$.

This says the number of successes in n Bernoulli trials is approximately Normal with parameters $\mu = np$ and $\sigma = \sqrt{np(1-p)}$.

Remark 2.31

(1) By convention we may apply the theorem if both $n\ p \geq 5$ and $n\ (1-p) \geq 5$. This choice for n is related to p and these two conditions would exclude small sample sizes and extreme values of p, i.e., $p \approx 0$ or $p \approx 1$.

(2) Since S_n is integer valued and normal random variables are not, we may use the **continuity correction** to get a better approximation. What that means is that for any integer x one should calculate $P(S_n \leq x)$ using $P(S_n \leq x + 0.5)$ and for $P(S_n < x)$ use $P(S_n \leq x - 0.5)$. That is,

$$P(S_n \leq x) \approx \text{normcdf}(-\infty, x + 0.5, np, \sqrt{np(1-p)})$$

$$P(S_n < x) \approx \text{normcdf}(-\infty, x - 0.5, np, \sqrt{np(1-p)}),$$

$$P(S_n \geq x) \approx \text{normcdf}(x - 0.5, \infty, np, \sqrt{np(1-p)}).$$

We may approximate $P(S_n = x) \approx P(x - 0.5 \leq S_n \leq x + 0.5) \approx \text{normcdf}(x - 0.5, x + 0.5, np, \sqrt{np(1-p)})$.

Example 2.32   Suppose you are going to play roulette 25 times, betting on red each time. What is the probability you win at least 14 games?

Remember that the probability of winning is $18/38 = p$. Let X be the number of games won. Then $X \sim \text{Binom}(25, 18/38)$ and what we want to find is $P(X \geq 14)$.

We may calculate this in two ways. First, using the binomial distribution

$$P(X \geq 14) = \sum_{x=14}^{25} \binom{25}{x}(18/38)^x(20/38)^{25-x} = 1 - \text{binomcdf}(25, 18/38, 13) = 0.2531.$$

This is the exact answer. Second, we may use the Central Limit Theorem which says

$$Z = \frac{S_{25} - 25\,(18/38)}{\sqrt{25\,(18/38)\,(20/38)}} \approx N(0, 1).$$

Consequently,

$$P(S_{25} \geq 14) = P\left(\frac{S_{25} - 25\,(18/38)}{\sqrt{25\,(18/38)\,(20/38)}} \geq \frac{14 - 25\,(18/38)}{\sqrt{25(18/38)\,(20/38)}}\right) \approx P(Z \geq 0.8644) = 0.1937.$$

This is not a great approximation. We can make it better by using the continuity correction. We have

$$P(S_{25} \geq 14) \approx \text{normcdf}(13.5, \infty, 25\,(18/38), \sqrt{25\,(18/38)\,(20/38)}) = 0.2533,$$

which is considerably better.

Figure 2.4 shows why the continuity correction gives a better estimate for a binomial.

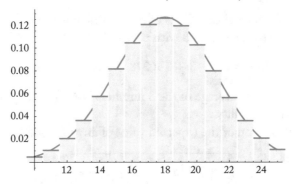

Figure 2.4: Normal approximation to binomial.

2.4 EXPECTATION, VARIANCE, MEDIANS, PERCENTILES

For any given random variable we are interested in basic properties like its mean value, median, the spread around the mean, etc. These are measures of central location of the rv and the spreads around these locations. These concepts are discussed here.

Definition 2.33   The **expected value** of a random variable X is

$$
E[X] = \begin{cases} \displaystyle\sum_{x} xP(X = x), & \text{if } X \text{ is discrete;} \\ \displaystyle\int_{-\infty}^{\infty} xf(x)\, dx, & \text{if } X \text{ is continuous with pdf } f. \end{cases}
$$

If $g : \mathbb{R} \to \mathbb{R}$ is a given function, the expected value of the rv $g(X)$ is[6]

$$
E[g(X)] = \begin{cases} \displaystyle\sum_{x} g(x)P(X = x), & \text{if } X \text{ is discrete;} \\ \displaystyle\int_{-\infty}^{\infty} g(x)f(x)\, dx, & \text{if } X \text{ is continuous with pdf } f. \end{cases}
$$

With this definition, you can see why it is frequently useful to write $E(g(X)) = \int g(x)P(X = x)\, dx$ even when X is a continuous rv. This abuses notation a lot and you have to keep in mind that $f(x) \neq P(X = x)$, which is zero when X is continuous.

From calculus we know that if we have a one-dimensional object with density $f(x)$ at each point, then $\int xf(x)\, dx = E(X)$ gives the **center of gravity** of the object. If X is discrete,

---

[6]We frequently write $E[g(X)] = E(g(X)) = Eg(X)$ and drop the braces or parentheses.

the expected value is an average of the values of X, weighted by the probability it takes on each value. For example, if X has values 1, 2, 3 with probabilities $1/8, 3/8, 1/2$, respectively, then

$$E[X] = 1 \times 1/8 + 2 \times 3/8 + 3 \times 1/2 = 19/8.$$

On the other hand, the straight average of the 3 numbers is 2. The straight average corresponds to each value with equal probability.

Now we have a definition of the expected value of any function of X. In particular,

$$\boxed{E[X^2] = \int_{-\infty}^{\infty} x^2 f(x)\, dx.}$$

We need this if we want to see how the random variable spreads its values around the mean.

Definition 2.34   The **variance** of a rv X is $Var(X) = E(X - (E[X]))^2$. Written out, the first step is to find the constant $\mu = E[X]$ and then

$$Var(X) = \begin{cases} \displaystyle\sum_x (x - \mu)^2 P(X = x), & \text{if } X \text{ is discrete;} \\ \displaystyle\int_{-\infty}^{\infty} (x - \mu)^2 f(x)\, dx, & \text{if } X \text{ is continuous.} \end{cases}$$

The **standard deviation**, abbreviated SD, of X, $SD(X) = \sqrt{Var(X)}$.

Another measure of the spread of a distribution is the median and the percentiles. Here's the definition.

Definition 2.35   The **median** $m = med(X)$ of a random variable X is defined to be the real number such that $P(X \le m) = P(X \ge m) = \frac{1}{2}$. The median is also known as the **50th percentile**.

Given a real number $0 < q < 1$, the $100q^{th}$ **percentile** of X is the number x_q such that $P(X \le x_q) = q$.

The **interquartile range** of a rv is $IQR = Q_3 - Q_1$, i.e., the 75th percentile minus the 25th percentile. Q_1 is the first quartile, the 25th percentile, and Q_3 is the third quartile, the 75th percentile. The median is also known as Q_2, the second quartile.

In other words, $100q\%$ of the values of X are **below** x_q. Percentiles apply to any random variable and give an idea of the shape of the density. Note that percentiles do not have to be unique, i.e., there may be several x_q's resulting in the same q.

Example 2.36   If $Z \sim N(0, 1)$ we may calculate $E[X] = \int_{-\infty}^{\infty} x \dfrac{1}{\sqrt{2\pi}} e^{-x^2/2}\, dx$ using substitution $(z = x^2/2)$ and obtain $E[Z] = 0$. Then we calculate $E[X^2] = \int_{-\infty}^{\infty} x^2 \dfrac{1}{\sqrt{2\pi}} e^{-x^2/2}\, dx$

using integration by parts. We get $E[Z^2] = 1$ and then $Var[Z] = E[Z^2] - (E[Z])^2 = 1$. **The parameters $\mu = 0$ and $\sigma = 1$ represent the mean and SD of Z.** In general, if $X \sim N(\mu, \sigma)$ we write $X = \sigma Z + \mu$ with $Z \sim N(0, 1)$, and we see that $E[X] = \sigma \times 0 + \mu = \mu$ and $Var[X] = \sigma^2 Var[Z] = \sigma^2$ so that $SD(X) = \sigma$.

Example 2.37   Suppose we know that LSAT scores follow a normal distribution with mean 155 and SD 13. You take the test and score 162. What percent of people taking the test did worse than you?

This is asking for $P(X \leq 162)$ knowing $X \sim N(155, 13)$. That's easy since $P(X \leq 162) =$ normalcdf$(-\infty, 162, 155, 13) = 0.704$. In other words, 162 is the 70.4 percentile of the scores.

Suppose instead someone told you that her score was in the 82nd percentile and you want to know her actual score. To find that, we are looking to solve $P(X \leq x_{.82}) = 0.82$. To find this using technology[7] we have $x_{.82} = $ invNorm$(.82, 155, 13) = 166.89$, so she scored about 167.

Now here's a proposition which says that the mean is the best estimate of a rv X in the mean square sense, and the median is the best estimate in the mean absolute deviation sense.

Proposition 2.38

(1) We have the alternate formula $Var(X) = E[X^2] - (E[X])^2$.

(2) The mean of X, $E[X] = \mu$ is the unique constant a which minimizes $E[X - a]^2$. Then $\min_a E[X - a]^2 = E[X - \mu]^2 = Var(X)$.

(3) A median $med(X)$ is a constant which provides a minimum for $E|X - a|$. In other words, $\min_a E|X - a| = E|X - med(X)|$.

The second statement says that the variance is the minimum of the mean squared distance of the rv X to its mean. The third statement says that a median (which may not be unique) satisfies a similar property for the absolute value of the distance.

Proof.   (1) Set $\mu = E[X]$

$$Var(X) = E(X - E[X])^2 = E\left[X^2 - 2X\mu + \mu^2\right]$$
$$= E\left[X^2\right] - 2\mu E[X] + \mu^2 = E\left[X^2\right] - \mu^2.$$

(2) We will assume X is a continuous rv with pdf f. Then, with
$G(a) = \int_{-\infty}^{\infty}(x - a)^2 f(x)\, dx$,

$$G'(a) = \frac{d}{da}\int_{-\infty}^{\infty}(x - a)^2 f(x)\, dx = \int_{-\infty}^{\infty} -2(x - a) f(x)\, dx = 0$$

[7]On a TI-84 the command is invNorm(q,mean,SD).

implies $\int x f(x)\, dx = \int a f(x)\, dx = a$. This assumes we can interchange the derivative and the integral. Furthermore, $G''(a) = 2 \int f(x)\, dx = 2 > 0$. Consequently, $a = E[X]$ provides a minimum for G. It is unique since G is strictly concave up.

(3) This is a little trickier since we can't take derivatives at first. We get rid of absolute value signs first.

$$E|X - a| = \int_{-\infty}^{\infty} |x - a| f(x)\, dx = \int_{-\infty}^{a} -(x - a) f(x)\, dx + \int_{a}^{\infty} (x - a) f(x)\, dx \equiv H(a).$$

Now we take derivatives

$$H'(a) = \int_{-\infty}^{a} f(x)\, dx - \int_{a}^{\infty} f(x)\, dx = 0 \implies \int_{-\infty}^{a} f(x)\, dx = \int_{a}^{\infty} f(x)\, dx \equiv \alpha.$$

Since $\int f(x)\, dx = 1 = \int_{-\infty}^{a} + \int_{a}^{\infty}, \implies \int_{-\infty}^{a} f(x)\, dx = 1 - \alpha$. But then $1 - \alpha = \alpha \implies \alpha = \frac{1}{2}$. We conclude that

$$P(X \le a) = \int_{-\infty}^{a} f(x)\, dx = \int_{a}^{\infty} f(x)\, dx = P(X \ge a) = \frac{1}{2},$$

and this says that a is a median of X. Furthermore, $H''(a) = 2 f(a) \ge 0$, so $a = med(X)$ does provide a minimum (but note that H is not necessarily strictly concave up).

$\square$

2.4.1   MOMENT-GENERATING FUNCTIONS

In the beginning of this section we defined the expected value of a function g of a rv X as $Eg(X) = \int g(x) f(x)\, dx$, where f is the pdf of X. We now consider a special and very useful function of X. This will give us a method of calculating means and variances usually in a much simpler way than doing it directly.

Definition 2.39   The moment-generating function (mgf) of a rv X is $M(t) = E[e^{tX}]$. Explicitly, we define

$$M(t) = \begin{cases} \displaystyle\int_{-\infty}^{\infty} e^{tx} f(x)\, dx, & \text{if } X \text{ is continuous;} \\[2ex] \displaystyle\sum_{x} e^{tx} P(X = x), & \text{if } X \text{ is discrete.} \end{cases}$$

We assume the integral or sum exists for all $t \in (-\delta, \delta)$ for some $\delta > 0$.

One reason the mgf is so useful is the following theorem. It says that if we know the mgf, we can find moments, i.e., $E(X^n), n = 1, 2, \dots,$ by taking derivatives.

Theorem 2.40   If X has the mgf $M(t)$, then $E[X^n] = \frac{d^n}{dt^n} M(t)\big|_{t=0}$.

Proof. The proof is easy if we assume that we can switch integral and derivatives.

$$\frac{d^n}{dt^n} M(t) = \int_{-\infty}^{\infty} \frac{d^n}{dt^n} e^{tx} f(x) \, dx = \int_{-\infty}^{\infty} x^n e^{tx} f(x) \, dx.$$

Plug in $t = 0$ in the last integral to see $\int_{-\infty}^{\infty} x^n e^{tx}|_{t=0} f(x) \, dx = \int_{-\infty}^{\infty} x^n f(x) \, dx = EX^n.$ □

Example 2.41 Let's use the mgf to find the mean and variance of $X \sim \text{Binom}(n, p)$.

$$M(t) = \sum_{x=0}^{n} e^{tx} P(X = x) = \sum_{x=0}^{n} e^{tx} \binom{n}{x} p^x (1 - p)^{n-x}$$

$$= \sum_{x=0}^{n} \binom{n}{x} (pe^t)^x (1 - p)^{n-x} = \boxed{(pe^t + (1 - p))^n}.$$

We used the Binomial Theorem from algebra[8] in the last line. Now that we know the mgf we can find any moment by taking derivatives. Here are the first two:

$$M'(t) = npe^t (pe^t + (1 - p))^{n-1} \implies M'(0) = EX = np,$$

and

$$M''(t) = n(n - 1)p^2 e^{2t} \left(pe^t + (1 - p)\right)^{n-2} + npe^t \left(pe^t + (1 - p)\right)^{n-1}$$
$$\implies EX^2 = M''(0) = n(n - 1)p^2 + np.$$

The variance is then $Var(X) = EX^2 - (EX)^2 = n(n - 1)p^2 + np - n^2 p^2 = np(1 - p).$

2.4.2 MEAN AND VARIANCE OF SOME IMPORTANT DISTRIBUTIONS

Now we use the mgf to calculate the mean and variances of some of the important continuous distributions.

(a) $\boxed{X \sim \text{Unif}[a, b]}$, $f(x) = \frac{1}{b-a}, a < x < b$. The mgf is

$$M(t) = \int_a^b e^{tx} \frac{1}{b - a} \, dx = \frac{1}{b - a} \frac{1}{t} e^{tx} \bigg|_a^b = \frac{e^{tb} - e^{ta}}{t(b - a)}.$$

Then

$$M'(t) = \frac{e^{at}(at - 1) + e^{bt}(1 - bt)}{(a - b)t^2} \quad \text{and} \quad \lim_{t \to 0} M'(t) = \frac{a + b}{2}.$$

---

[8]$(a + b)^n = \sum_{k=0}^{n} \binom{n}{k} a^k b^{n-k}.$

We conclude $EX = \frac{a+b}{2}$. While we could find $M''(0) = EX^2$, it is actually easier to find this directly.

$$EX^2 = \int_a^b x^2 \frac{1}{b-a} \, dx = \frac{b^3 - a^3}{3(b-a)} \implies Var(X) = EX^2 - (EX)^2 = \frac{(b-a)^2}{12}.$$

(b) $\boxed{X \sim \text{Exp}(\lambda)}$, $f(x) = \lambda e^{-\lambda x}$, $x > 0$. We get

$$M(t) = E[e^{tX}] = \int_0^\infty e^{tx} \lambda e^{-\lambda x} \, dx = \frac{\lambda}{\lambda - t}, \quad \text{if } t < \lambda.$$

$$M'(t) = \frac{\lambda}{(\lambda - t)^2}, \quad M'(0) = \frac{1}{\lambda} = E[X], \text{ and } M''(t) = \frac{2\lambda}{(\lambda - t)^3}, \quad M''[0] = \frac{2}{\lambda^2} = E[X^2]$$

$$Var(X) = EX^2 - (EX)^2 = \frac{2}{\lambda^2} - \frac{1}{\lambda^2} = \frac{1}{\lambda^2}.$$

(c) $\boxed{X \sim N(0, 1)}$, $f(x) = \frac{1}{\sqrt{2\pi}} e^{-\frac{x^2}{2}}$, $-\infty < x < \infty$. The mgf for the standard normal distribution is

$$M(t) = \int_{-\infty}^\infty e^{tx} \frac{1}{\sqrt{2\pi}} e^{-\frac{x^2}{2}} \, dx = \frac{1}{\sqrt{2\pi}} \int_{-\infty}^\infty e^{tx - \frac{x^2}{2}} \, dx$$

$$= \frac{1}{\sqrt{2\pi}} \int_{-\infty}^\infty e^{t^2/2} e^{-\frac{1}{2}(x-t)^2} \, dx = e^{t^2/2} \frac{1}{\sqrt{2\pi}} \int_{-\infty}^\infty e^{-\frac{1}{2}(x-t)^2} \, dx = e^{t^2/2}$$

since $\frac{1}{\sqrt{2\pi}} \int_{-\infty}^\infty e^{-\frac{1}{2}(x-t)^2} \, dx = 1$. Now we can find the moments fairly simply.

$$M'(t) = e^{\frac{t^2}{2}} t \implies M'(0) = EX = 0 \text{ and } M''(t) = e^{\frac{t^2}{2}} t^2 + e^{\frac{t^2}{2}} \implies M''(0) = EX^2 = 1$$

$$Var(X) = EX^2 - (EX)^2 = 1.$$

(d) $\boxed{X \sim N(\mu, \sigma)}$. All we have to do is convert X to standard normal. Let $Z = \frac{X - \mu}{\sigma}$. We know $Z \sim N(0, 1)$ and we may use the previous part to write $M_Z(t) = e^{t^2/2}$. How do we get the mgf for X from that? Well, we know $X = \sigma Z + \mu$ and so

$$M_X(t) = E e^{tX} = E e^{(\sigma Z + \mu)t} = e^{\mu t} E e^{(t\sigma)Z} = e^{\mu t} e^{\frac{1}{2}(\sigma t)^2} = e^{\mu t + \frac{1}{2} \sigma^2 t^2}.$$

Then $M'(t) = e^{\frac{\sigma^2 t^2}{2} + \mu t} \left(\mu + \sigma^2 t \right)$ so that $M'(0) = EX = \mu$. Next

$$M''(t) = e^{\frac{\sigma^2 t^2}{2} + \mu t} \left(\sigma^2 + \left(\mu + \sigma^2 t \right)^2 \right) \implies M''(0) = \sigma^2 + \mu^2.$$

This gives us $Var(X) = EX^2 - (EX)^2 = (\sigma^2 + \mu^2) - \mu^2 = \sigma^2$.

Later we will need the following results. The first part says that if two rvs have the same mgf, then they have the same distribution. The second part says that if the mgfs of a sequence of rvs converges to an mgf, then the cdfs must also converge to the cdf of the limit rv.

Theorem 2.42

(1) If X and Y are two rvs such that $M_X(t) = M_Y(t)$ (for all t close to 0), then X and Y have the same cdfs.

(2) If $X_k, k = 1, 2, \ldots$, is a sequence of rvs with mgf $M_k(t), k = 1, 2, \ldots$, and cdf $F_k(x), k = 1, 2, \ldots$, respectively, and if $\lim_{k \to \infty} M_k(t) = M_X(t)$ and $M_X(t)$ is an mgf then there is a unique cdf F_X and $\lim_{k \to \infty} F_k(x) = F_X(x)$ at each x a point of continuity of F_X.

Example 2.43   An rv X has density $f(x) = \frac{1}{\sqrt{2\pi}} \frac{1}{\sqrt{x}} e^{-\frac{x}{2}}, x > 0$ and $f(x) = 0$ if $x < 0$. First, we find the mgf of X.

$$M_X(t) = \frac{1}{\sqrt{2\pi}} \int_0^\infty e^{tx} \frac{1}{\sqrt{x}} e^{-\frac{x}{2}} \, dx$$

$$= \frac{2}{\sqrt{2\pi}} \int_0^\infty e^{u^2(t-\frac{1}{2})} \, du \quad \text{setting } u = \sqrt{x}$$

$$= \frac{2}{\sqrt{2\pi}} \frac{1}{\sqrt{1-2t}} \int_0^\infty e^{-z^2/2} \, dz \quad \text{setting } z = u\sqrt{2}\sqrt{1-2t}, \; t < \frac{1}{2}$$

$$= \frac{1}{\sqrt{1-2t}}, \quad \text{for } t < \frac{1}{2}, \text{ since } \frac{2}{\sqrt{2\pi}} \int_0^\infty e^{-z^2/2} \, dz = 1.$$

Then,

$$EX = M_X'(0) = 1 \quad \text{and} \quad EX^2 = M_X''(0) = 3 \implies Var(X) = 3 - 1^2 = 2.$$

Where does this density come from? To answer this, let $Z \sim N(0, 1)$ and let's find the pdf of $Y = Z^2$.

$$F_Y(y) = P(Z^2 \le y) = P(-\sqrt{y} \le Z \le \sqrt{y}) = \frac{1}{\sqrt{2\pi}} \int_{-\sqrt{y}}^{\sqrt{y}} e^{-x^2/2} \, dx$$

$$f(y) = F_Y'(y) = \frac{1}{\sqrt{2\pi}} \left(e^{-y/2} \frac{1}{2\sqrt{y}} - e^{-y/2} \frac{-1}{2\sqrt{y}} \right)$$

$$= \frac{1}{\sqrt{2\pi}} \left(e^{-y/2} \frac{2}{2\sqrt{y}} \right), \quad y > 0.$$

If $y \leq 0$, $f(y) = 0$. This shows that the density we started with is the density for Z^2. Now we calculate the mgf for Z^2.

$$M_{Z^2}(t) = E[e^{tZ^2}] = \frac{1}{\sqrt{2\pi}} \int_{-\infty}^{\infty} e^{tz^2} e^{-z^2/2} \, dz, \text{ since } Z \sim N(0,1)$$

$$= \frac{1}{\sqrt{2\pi}} \int_{-\infty}^{\infty} e^{z^2(t - \frac{1}{2})} \, dz = M_X(t)$$

since if we compare the last integral with the second integral in the computation of M_X we see that they are the same. This means that $M_X(t) = M_{Z^2}(t)$ and part (1) of the Theorem 2.42 says that X and Z^2 must have the same distribution.

Definition 2.44   The rv $X = Z^2$ with density given by $f(x) = \frac{1}{\sqrt{2\pi}} \frac{1}{\sqrt{x}} e^{-\frac{x}{2}}$, $x > 0$ and $f(x) = 0$ if $x < 0$, is called χ^2 with 1 degree of freedom, written as $X \sim \chi^2(1)$.

Remark 2.45   We record here the mean and variance of some important discrete distributions.

(a) $X \sim \text{Binom}(n, p)$, $EX = np$, $Var(X) = np(1 - p)$.

(b) $X \sim \text{Geom}(p)$, $EX = \dfrac{1}{p}$, $Var(X) = \dfrac{1 - p}{p^2}$, $M_X(t) = \dfrac{pe^t}{1 - (1 - p)e^t}$.

(c) $X \sim \text{HyperGeom}(N, r, n)$, $EX = np$, $p = r/N$, $Var(X) = np(1 - p)\dfrac{N - n}{N - 1}$.

(d) $X \sim \text{NegBinom}(r, p)$, $EX = r\dfrac{1 - p}{p}$, $Var(X) = r\dfrac{1 - p}{p^2}$. $M_X(t) = \left(\dfrac{pe^t}{1 - (1 - p)e^t}\right)^r$.

(e) $X \sim \text{Poisson}(\lambda)$, $EX = \lambda$, $Var(X) = \lambda$, $M_X(t) = e^{\lambda(e^t - 1)}$.

2.5   JOINT DISTRIBUTIONS

In probability and statistics we are often confronted with a problem involving more than one random variable which may or may not depend on each other. We have to study jointly distributed random variables if we want to calculate things like $P(X + Y \leq w)$.

Definition 2.46

(1) If X and Y are two random variables, the joint cdf is $F_{X,Y}(x, y) = P(\{X \leq x\} \cap \{Y \leq y\})$. In general, we write this as $F_{X,Y}(x, y) = P(X \leq x, Y \leq y)$.

(2) If X and Y are discrete, the pmf of (X, Y) is $p(x, y) = P(X = x, Y = y)$.

(3) A joint density function is a function $f_{X,Y}(x, y) \geq 0$ with $\int_{-\infty}^{\infty} \int_{-\infty}^{\infty} f_{X,Y}(x, y) \, dx \, dy = 1$. The pair of rvs (X, Y) is continuous if there is a joint density function and then

$$F_{X,Y}(x, y) = \int_{-\infty}^{x} \int_{-\infty}^{y} f_{X,Y}(u, v) \, du \, dv.$$

(4) If we know $F_{X,Y}(x, y)$ then the joint density is $f_{X,Y}(x, y) = \frac{\partial^2 F_{X,Y}(x,y)}{\partial x \partial y}$.

Knowing the joint distribution of X and Y means we have full knowledge of X and Y individually. For example, if we know $F_{X,Y}(x, y)$, then

$$F_X(x) = F_{X,Y}(x, \infty) = \lim_{y \to \infty} F_{X,Y}(x, y), \quad F_Y(y) = F_{X,Y}(\infty, y).$$

The resulting F_X and F_Y are called the **marginal cumulative distribution functions**. The **marginal densities** when there is a joint density are given by

$$f_X(x) = \int_{-\infty}^{\infty} f_{X,Y}(x, y) \, dy \quad \text{and} \quad f_Y(y) = \int_{-\infty}^{\infty} f_{X,Y}(x, y) \, dx.$$

Example 2.47   The function $f(x, y) = \begin{cases} 8xy, & 0 \leq x < y \leq 1, \\ 0 & \text{otherwise.} \end{cases}$ is given. First we verify it is a joint density. Since $f \geq 0$ all we need to check is that the double integral is one.

$$\int_{-\infty}^{\infty} \int_{-\infty}^{\infty} f(x, y) \, dx \, dy = \int_0^1 \int_0^y 8xy \, dx \, dy = \int_0^1 8y \left(\frac{1}{2} y^2 \right) dy = 1.$$

To find the marginal densities we have

$$f_X(x) = \int_{-\infty}^{\infty} f(x, y) \, dy = \begin{cases} \int_x^1 8xy \, dy = 4x(1 - x^2), & \text{if } 0 \leq x \leq 1 \\ 0, \text{ otherwise.} \end{cases}$$

$$f_Y(y) = \int_{-\infty}^{\infty} f(x, y) \, dx = \begin{cases} \int_0^y 8xy \, dx = 4y^3, & \text{if } 0 \leq y \leq 1 \\ 0, \text{ otherwise.} \end{cases}$$

If X and Y are discrete rvs, the joint pmf is $p(x, y) = P(X = x, Y = y)$. The marginals are then given by $p_X(x) = P(X = x) = \sum_y p(x, y)$ and $p_Y(y) = P(Y = y) = \sum_x p(x, y)$.

In general, to find the probability that for any set $C \subset \mathbb{R} \times \mathbb{R}$, the pair $(X, Y) \in C$ has probability defined by

$$P((X, Y) \in C) = \begin{cases} \iint\limits_C f_{X,Y}(x, y)\, dxdy, & \text{if } X, Y \text{ are continuous,} \\ \sum\sum\limits_{(x,y)\in C} p_{X,Y}(x, y), & \text{if } X, Y \text{ are discrete.} \end{cases} \qquad (2.1)$$

We also have expected values of functions of rvs.

Definition 2.48   If (X, Y) have joint density $f_{X,Y}(x, y)$, the expected value of a function of the rvs is

$$E[g(X, Y)] = \begin{cases} \displaystyle\int_{-\infty}^{\infty}\int_{-\infty}^{\infty} g(x, y)\, f_{X,Y}(x, y)\, dx\, dy, & \text{if } X, Y \text{ are continuous,} \\ \displaystyle\sum_{x,y} g(x, y) P(X = x, Y = y), & \text{if } X, Y \text{ are discrete.} \end{cases}$$

Example 2.49   We calculate $E(X + Y)$ assuming we have the joint density of (X, Y) given by $f_{X,Y}(x, y)$. By definition

$$E(X + Y) = \iint (x + y) f_{X,Y}(x, y)\, dxdy$$

$$= \iint x f_{X,Y}(x, y)\, dxdy + \int\int y f_{X,Y}(x, y)\, dxdy$$

$$= \int x \left(\int f_{X,Y}(x, y)\, dy \right) dx + \int y \left(\int y f_{X,Y}(x, y)\, dx \right) dy$$

$$= \int x f_X(x)\, dx + \int y f_Y(y)\, dy = E(X) + E(Y).$$

Notice that the first E uses the joint density $f_{X,Y}$ while the second and third E's use f_X and f_Y, respectively.

Example 2.50   Suppose (X, Y) have joint density $f(x, y) = 1, 0 \leq x, y \leq 1$, and $f(x, y) = 0$ otherwise. This models picking a random point (x, y) in the unit square. If we want to calculate $P(X < Y)$, this uses the density.

$$P(X < Y) = \iint\limits_{0 \leq x < y \leq 1} f(x, y)\, dx\, dy = \int_0^1 \int_0^y 1\, dx\, dy = \left. \frac{y^2}{2} \right|_0^1 = \frac{1}{2}.$$

Similarly, we may calculate

$$P\left(X^2 + Y^2 \le \frac{1}{4}\right) = \iint\limits_{0 \le x^2 + y^2 \le \frac{1}{4}} 1 \, dx \, dy = \text{area of semicicle in square} = \frac{\pi}{16}.$$

Also,

$$E\left(X^2 + Y^2\right) = \int_0^1 \int_0^1 \left(x^2 + y^2\right) \times f(x, y) \, dx \, dy = \frac{2}{3}.$$

Remark 2.51   In general, if we are given a set $D \subset \mathbb{R}^2$ the density

$$f(x, y) = \begin{cases} \dfrac{1}{\text{area of } D}, & (x, y) \in D, \\ 0, & \text{otherwise,} \end{cases}$$

is called a uniform density on D and the rvs $(X, Y) \sim \text{Unif}(D)$.

Example 2.52   For the rvs (X, Y) with joint density $f(x, y) = \begin{cases} 8xy, & 0 \le x < y \le 1, \\ 0 & \text{otherwise,} \end{cases}$ we'll
find $E(X + Y)$ and $E(XY)$.

$$E(X + Y) = \int_0^1 \int_0^y 8xy(x + y) \, dx \, dy = \frac{4}{3} \quad \text{and} \quad E(XY) = \int_0^1 \int_x^1 8xy(xy) \, dy \, dx = \frac{4}{9}.$$

Note that

$$E(X) = \int_0^1 4x\left(1 - x^2\right) x \, dx = \frac{8}{15} \quad \text{and} \quad E(Y) = \int_0^1 4y^3 y \, dy = \frac{4}{5},$$

so that $E(X + Y) = E(X) + E(Y)$ but $E(XY) \ne E(X) \times E(Y)$.

You see that $E(XY) \ne E(X) \times E(Y)$ in general, but there are important cases when this is true. For that we need the notion of independent random variables.

2.5.1   INDEPENDENT RANDOM VARIABLES

Independence of random variables, which has the intuitive meaning that one random variable doesn't affect the other, is a central idea in probability.

Definition 2.53   Random variables X, Y are independent if

$$P\left(X \le x, Y \le y\right) = P\left(X \le x\right) P\left(Y \le y\right), \ \forall \, x \in \mathbb{R}, y \in \mathbb{R}.$$

If (X, Y) has a joint density $f_{X,Y}$, X has density f_X, and Y has density f_Y, then independence means that the joint density factors into the individual densities:

$$f_{X,Y}(x, y) = f_X(x) f_Y(y).$$

One of the main consequences of independence is the following fact. It says the expected value of a product of rvs is the product of the expected value of each rv.

Proposition 2.54   If X, Y are independent

$$E[XY] = E[X] \times E[Y].$$

In fact, for any functions g, h, we have $E[g(X)h(Y)] = E[g(X)] \times E[h(Y)]$.

Proof. By definition,

$$E[XY] = \int_{-\infty}^{\infty} \int_{-\infty}^{\infty} xy \, f_{X,Y}(x, y) \, dx \, dy = \int_{-\infty}^{\infty} \int_{-\infty}^{\infty} xy \, f_X(x) f_Y(y) \, dx \, dy$$

$$= \int_{-\infty}^{\infty} x \, f_X(x) \, dx \times \int_{-\infty}^{\infty} y \, f_Y(y) \, dy = E[X] \times E[Y].$$

The proof of the second statement is almost identical.   □

Independence also allows us to find an explicit expression for the joint cumulative distribution of the sum of two random variables.

Proposition 2.55   If X, Y are independent continuous rvs then

$$F_{X+Y}(w) = P(X + Y \leq w) = \int_{-\infty}^{\infty} P(X \leq w - y) f_Y(y) \, dy.$$

This is really another application of the Law of Total Probability. To see this

$$P(X + Y \leq w) = \int P(X + Y \leq w, Y = y) \, dy$$

$$= \int P(X \leq w - y) P(Y = y) \, dy = \int F_X(w - y) f_Y(y) \, dy.$$

The first equality uses the Law of Total Probability and the second equality uses the independence.

Example 2.56   Suppose X and Y are independent $\text{Exp}(\lambda)$ rvs. Then, for $w \geq 0$,

$$P(X + Y \leq w) = \int_0^\infty F_X(w - y) f_Y(y) \, dy = \int_0^w \left(1 - e^{-\lambda(w-y)}\right) \lambda e^{-\lambda y} \, dy$$

$$= 1 - (\lambda w + 1)e^{-\lambda w} = F_{X+Y}(w).$$

If $w < 0$, $F_{X+Y}(w) = 0$. To find the density we take the derivative with respect to w to get

$$f_{X+Y}(w) = \lambda^2 w \, e^{-\lambda w}, \quad w \geq 0.$$

It turns out that this is the pdf of a so-called Gamma $(\lambda, 2)$ rv.

2.5.2   COVARIANCE AND CORRELATION

A very important quantity measuring the linear relationship between two rvs is the following.

Definition 2.57   Given two rvs X, Y, the **covariance** of X, Y is defined by

$$\boxed{Cov(X, Y) = E[XY] - E[X]E[Y] = E[(X - EX)] \, E[(Y - EY)].}$$

The **correlation coefficient** is defined by

$$\boxed{\rho(X, Y) = \frac{Cov(X, Y)}{\sigma_X \, \sigma_Y}, \quad \sigma_X^2 = Var(X), \ \sigma_Y^2 = Var(Y).}$$

X and Y are said to be **uncorrelated** if $Cov(X, Y) = 0$ or, equivalently, $\rho(X, Y) = 0$.

It looks like covariance measures how independent X and Y are. It is certainly true that if X, Y are independent, then $\rho(X, Y) = 0$, but the reverse is false.

Example 2.58   Suppose X, Y have the joint pmf $P(X = -1, Y = 0) = P(X = 0, Y = -1) = P(X = 0, Y = 1) = P(X = 1, Y = 0) = \frac{1}{4}$. For all other cases $P(X = i, Y = j) = 0$. As defined earlier, given the joint density $P(X = x, Y = y)$,

$$P(X = x) = \sum_y P(X = x, Y = y) \ \text{ and } \ P(Y = y) = \sum_x P(X = x, Y = y)$$

are the **marginals** of X, Y. In this example $P(X = -1) = \frac{1}{4}$, $P(X = 0) = \frac{1}{2}$, and $P(X = 1) = \frac{1}{4}$. Similarly, $P(Y = -1) = \frac{1}{4}$, $P(Y = 0) = \frac{1}{2}$, $P(Y = 1) = \frac{1}{4}$. We can arrange all this in a two-way table

X	Y			
	−1	0	1	$P(X = x)$
−1	0	1/4	0	1/4
0	1/4	0	1/4	1/2
1	0	1/4	0	1/4
$P(Y = y)$	1/4	1/2	1/4	1

The sum of each row is $P(X = x)$ while the sum of each column is $P(Y = y)$. Each element of the matrix is $P(X = x, Y = y)$. If X, Y are independent, the (x, y) element of the matrix must be the product of the marginals, i.e., $P(X = x, Y = y) = P(X = x)P(Y = y)$. You can see from the table that is not true so X, Y are not independent. On the other hand,

$$E[XY] = \sum_{x,y} x\, yP(X = x, Y = y) = 0$$

$$E[X] = (-1)\frac{1}{4} + (+1)\frac{1}{4} = 0, \quad \text{and} \quad E[Y] = (-1)\frac{1}{4} + (+1)\frac{1}{4} = 0.$$

which means $Cov(X, Y) = 0$ and so X, Y are uncorrelated.

Here's one of the more important implications of independence.

Theorem 2.59   If X, Y are rvs $\boxed{Var(X + Y) = Var(X) + 2Cov(X, Y) + Var(Y)}$. If X, Y are uncorrelated, then $Var(X + Y) = Var(X) + Var(Y)$.

Proof. This is a calculation.

$$\begin{aligned} Var(X + Y) &= E(X + Y)^2 - (EX + EY)^2 \\ &= E(X^2 + 2XY + Y^2) - (EX)^2 - 2EXEY - (EY)^2 \\ &= Var(X) + 2Cov(X, Y) + Var(Y). \end{aligned}$$

If X, Y are uncorrelated, $Cov(X, Y) = 0$. □

Remark 2.60   This can be extended to n rvs $X_1, \ldots, X_n$. If they are uncorrelated (which is true if they are independent), $Var(X_1 + \cdots + X_n) = Var(X_1) + \cdots + Var(X_n)$.

2.5.3   THE GENERAL CENTRAL LIMIT THEOREM

For statistics, one of the major applications of independence is the following fact:

$$X_i \sim N(\mu_i, \sigma_i), i = 1, 2, \ldots, n, \text{ and independent} \implies \sum_{i=1}^{n} X_i \sim N\left(\sum_{i=1}^{n} \mu_i, \sqrt{\sum_{i=1}^{n} \sigma_i^2}\right).$$

We can see this using mgfs and the following proposition.

Proposition 2.61   Let $X_1, X_2, \ldots, X_n$ be independent rvs with mgf $M_{X_i}(t), i = 1, 2, \ldots, n$. Let $S_n = X_1 + \cdots + X_n$. Then $M_{S_n}(t) = M_{X_1}(t) \cdot M_{X_2}(t) \cdots M_{X_n}(t)$.

This is directly from the definition and the independence. In fact,

$$M_{S_n}(t) = Ee^{(X_1 + \cdots + X_n)t} = Ee^{tX_1} \cdots Ee^{tX_n}.$$

Therefore, if $X_i \sim N(\mu_i, \sigma_i), i = 1, 2, \ldots, n$, and they are independent, we have

$$M_{S_n}(t) = \prod_{i=1}^{n} \exp\left\{ t\mu_i + \frac{1}{2}\sigma_i^2 t^2 \right\} = \exp\left\{ t \sum \mu_i + \frac{t^2}{2} \sum \sigma_i^2 \right\}.$$

Since mgfs determine a distribution uniquely according to Theorem 2.42, we see that $S_n \sim N\left(\sum \mu_i, \sqrt{\sum \sigma_i^2} \right)$.

Example 2.62   The sum of independent Geom(p) random variables is Negative Binomial. In particular, suppose X is the number of Bernoulli trials until we get r successes with probability p of success on each trial. Then $X = X_1 + X_2 + \cdots + X_r$, where $X_i \sim$ Geom$(p), i = 1, 2, \ldots, r$, is the number of trials until the first success. This is true since once we have a success we simply start counting anew from the last success until we get another success. Now, we have by independence,

$$E[X] = \sum_{i=1}^{r} E[X_i] = \frac{r}{p}, \quad \text{and} \quad Var[X] = \sum_{i=1}^{r} Var[X_i] = \frac{r(1-p)}{p^2}.$$

In addition, using the mgf of Geom(p), namely, $M_{X_i}(t) = \frac{e^t p}{1 - e^t(1-p)}, t < -\ln(1-p)$ we have

$$M_X(t) = \prod_{i=1}^{r} M_{X_i}(t) = \frac{e^{rt} p^r}{(1 - e^t(1-p))^r}, \quad t < -\ln(1-p).$$

and this must be the mgf of a Negative Binomial rv.

We have seen that the sum of independent normal rvs is exactly normal. The Central Limit Theorem says that even if the X_i's are not normal, the sum is approximately normal if the number of rvs is large. We have already seen the special case of this for Binomials but it is true in much more generality. The full proof is covered in more advanced courses.

Theorem 2.63   Central Limit Theorem. Let $X_1, X_2, \ldots$ be a sequence of independent rvs all having the same distributions and $EX_1 = \mu, Var(X_1) = \sigma^2$. Then for any $a, b \in \mathbb{R}$,

$$\lim_{n \to \infty} P\left(a \leq \frac{X_1 + \cdots + X_n - n\mu}{\sigma \sqrt{n}} \leq b \right) = P(a \leq Z \leq b),$$

where $Z \sim N(0, 1)$.

In other words, for large n, (generally $n \geq 30$),

$$\boxed{S_n = X_1 + \cdots + X_n \approx N(n\mu, \sigma \sqrt{n})}$$

and, dividing by n, since $E\frac{S_n}{n} = \frac{n\mu}{n} = \mu$, $Var(\frac{S_n}{n}) = \frac{1}{n^2}\sigma^2 n = \frac{\sigma^2}{n}$,

$$\boxed{\overline{X} \equiv \frac{S_n}{n} \approx N\left(\mu, \frac{\sigma}{\sqrt{n}}\right).}$$

This is true no matter what the distributions of the individual X_i's are as long as they all have the same finite means and variances.

Sketch of Proof of the CLT (Optional): We may assume $\mu = 0$ (why?) and we also may assume $a = -\infty$. Set $Z_n = S_n/(\sigma \sqrt{n})$. Then, if $M(t) = M_{X_i}(t)$ is the common mgf of the rvs X_i,

$$M_{Z_n}(t) = \left[M\left(\frac{t}{\sigma \sqrt{n}}\right)\right]^n = \exp(n \ln M(t/(\sigma \sqrt{n}))).$$

If we can show that

$$\lim_{n \to \infty} M_{Z_n}(t) = e^{t^2/2}$$

then by Theorem 2.42 we can conclude that the cdf of Z_n will converge to the cdf of the random variable that has mgf $e^{t^2/2}$. But that random variable is $Z \sim N(0, 1)$. That will complete the proof. Therefore, all we need to do is to show that

$$\lim_{n \to \infty} n \ln M(t/(\sigma \sqrt{n})) = \frac{t^2}{2}.$$

To see this, change variables to $x = t/(\sigma \sqrt{n})$ so that

$$\lim_{n \to \infty} n \ln M(t/(\sigma \sqrt{n})) = \lim_{x \to 0} \frac{t^2}{\sigma^2} \frac{\ln M(x)}{x^2}.$$

Since $\ln M(0) = 0$ we may use L'Hopital's rule to evaluate the limit. We get

$$\lim_{x \to 0} \frac{t^2}{\sigma^2} \frac{\ln M(x)}{x^2} = \frac{t^2}{\sigma^2} \lim_{x \to 0} \frac{M'(x)/M(x)}{2x}$$

$$= \frac{t^2}{2\sigma^2} \lim_{x \to 0} \frac{M''(x)}{xM'(x) + M(x)} \quad \text{using L'Hopital again}$$

$$= \frac{t^2}{2\sigma^2} \frac{M''(0)}{0\, M'(0) + M(0)} = \frac{t^2}{2\sigma^2} \frac{\sigma^2}{0 + 1} = \frac{t^2}{2},$$

since $M(0) = 1, M'(0) = EX = 0, M''(0) = EX^2 = \sigma^2$. This completes the proof. □

Example 2.64 Suppose an elevator is designed to hold 2000 pounds. The mean weight of a person getting on the elevator is 175 with standard deviation 15 pounds. How many people can board the elevator so that the chance it is overloaded is 1%?

Let $W = X_1 + \cdots + X_n$ be the total weight of n people who board the elevator. We don't know the distribution of the weights of individual people (which is probably not normal), but we do know $EX = 175$ and $Var(X) = 15^2$. By the central limit theorem, $W \approx N(175n, 15\sqrt{n})$ and we want to find n so that

$$P(W > 2000) = 0.01.$$

If we standardize W we get

$$0.01 = P(W > 2000) = P\left(\frac{W - 175n}{15\sqrt{n}} > \frac{2000 - 175n}{15\sqrt{n}}\right) = P\left(Z > \frac{2000 - 175n}{15\sqrt{n}}\right).$$

Using a calculator, we get $P(Z > z) = 0.01 \implies z = \text{invNorm}(0.99) = 2.326$. Therefore, it must be true that

$$\frac{2000 - 175n}{15\sqrt{n}} \geq 2.326 \implies n \leq 11.$$

The maximum number of people that can board the elevator and meet the criterion is 11. Without knowing the distribution of the weight of people, there is no other way to do this problem.

2.5.4 CHEBYCHEV'S INEQUALITY AND THE WEAK LAW OF LARGE NUMBERS

Suppose we have a rv X which has an arbitrary distribution but a finite mean $\mu = EX$ and variance $\sigma^2 = Var(X)$. Chebychev's inequality gives an upper bound on the chances X differs from it's mean without knowing anything about the distribution of X at all. Here's the inequality.

$$P(|X - \mu| \geq c) \leq \frac{\sigma^2}{c}, \quad \text{for any constant } c > 0. \tag{2.2}$$

The larger c is the smaller the probability can be. The argument for Chebychev is simple. Assume X has pdf f. Then

$$\sigma^2 = E|X - \mu|^2 = \int_{|x-\mu| \geq c} |x - \mu|^2 f(x)\, dx + \int_{|x-\mu| \leq c} |x - \mu|^2 f(x)\, dx$$

$$\geq \int_{|x-\mu| \geq c} |x - \mu|^2 f(x)\, dx \geq c^2 \int_{|x-\mu| \geq c} f(x)\, dx = c^2 P(|X - \mu| \geq c).$$

□

Chebychev is used to give us the Weak Law of Large Numbers which tells us that the mean of a random sample should converge to the population mean as the sample size goes to infinity.

Theorem 2.65 Weak Law of Large Numbers. Let $X_1, \ldots, X_n$ be a random sample, i.e., independent and all having the same distribution as the rv X which has finite mean $EX = \mu$ and finite variance $\sigma^2 = Var(X)$. Then, for any constant $c > 0$, with $\overline{X} = \frac{X_1 + \cdots + X_n}{n}$,

$$\lim_{n \to \infty} P\left(|\overline{X} - \mu| \geq c\right) = 0.$$

Proof. We know $E\overline{X} = \mu$ and $Var(\overline{X}) = \frac{\sigma^2}{n}$. By Chebychev's inequality,

$$P\left(|\overline{X} - \mu| \geq c\right) \leq \frac{Var(\overline{X})}{c} = \frac{\sigma^2}{nc} \to 0 \quad \text{as } n \to \infty.$$

□

The Strong Law of Large Numbers, which is beyond the scope of this book, says $\frac{X}{n} \to \mu$ in a much stronger way than the Weak Law says, so we are comfortable that the sample means do converge to the population mean.

2.6 $\chi^2(k)$, STUDENT'S t- AND F-DISTRIBUTIONS

In this section we will record three of the most important distributions for statistics.

2.6.1 $\chi^2(k)$ DISTRIBUTION

This is known as the χ^2 distribution with k degrees of freedom. We already encountered the $\chi^2(1) = Z^2$ distribution where we showed it is the same as a standard normal squared. There is a similar characterization for $\chi^2(k)$. In fact, let $Z_1, \ldots, Z_k$ be k independent $N(0, 1)$ rvs. Define

$$Y = Z_1^2 + Z_2^2 + \cdots + Z_k^2.$$

Then $Y \sim \chi^2(k)$. That is, a $\chi^2(k)$ rv is the sum of the squares of k independent normal rvs. In fact, if we look at the mgf of Y, we have using Example 2.43 and independence,

$$M_Y(t) = \prod_{i=1}^{k} M_{Z_i^2}(t) = \prod_{i=1}^{k} \frac{1}{\sqrt{1 - 2t}} = \left(\frac{1}{1 - 2t}\right)^{k/2}, \quad t < \frac{1}{2},$$

which is the mgf of a $\chi^2(k)$ rv which may be derived directly from the density. From, the mgf it is easy to see that $EY = k$ and $Var(Y) = 2k$. The main properties of Y are the following.

Remark 2.66   If $X \sim \chi^2(n), Y \sim \chi^2(m)$ and X, Y are independent, then $X + Y \sim \chi^2(n + m)$. To see why,

$$M_{X+Y}(t) = Ee^{t(X+Y)} = M_X(t)M_Y(t) = \left(\frac{1}{1-2t}\right)^{n/2} \left(\frac{1}{1-2t}\right)^{m/2} = \left(\frac{1}{1-2t}\right)^{(n+m)/2}$$

for $t < \frac{1}{2}$. Since distributions are uniquely determined by their mgf and the mgf of $X + Y$ is the mgf of a $\chi^2(n + m)$ rv, we know that $X + Y \sim \chi^2(n + m)$.

Remark 2.67   The $\chi^2(n)$ distribution is not symmetric. Therefore, if we want to find a, b so that $P(a < \chi^2(n) < b) = 1 - \alpha$ for some given $0 < \alpha < 1$, we set it up so the area to the right of b is $\frac{\alpha}{2}$ and the area to the left of a is also $\frac{\alpha}{2}$. Using a TI-8x calculator, the command is $a =$ invchi$(n, 1 - \alpha/2)$ and $b =$ invchi$(n, \alpha/2)$ where the first parameter is the area desired to the right of a or b. The program to get this is based on Newton's method for solving χ^2cdf$(0, x, n) = 1 - \alpha$ for x.

(1) Input "RT TAIL," A

(2) Input "D of F," N

(3) N $\rightarrow$ X

(4) For (J,1,9)

(5) X-(χ^2 cdf(0,X,N)+A-1)/χ^2 pdf(X,N) $\rightarrow$ X

(6) End

(7) Disp X

(8) Stop

2.6.2   STUDENT'S t-DISTRIBUTION

This is a combination of two independent rvs, a $N(0, 1)$ rv and a $\chi^2(k)$ rv which arises naturally in statistics.

$$T(k) = \frac{Z}{\sqrt{\chi^2(k)/k}}, \quad Z \sim N(0, 1). \tag{2.3}$$

We say that T has a **Student's t-distribution** with k degrees of freedom.

Remark 2.68   We will come to this later but this will come from looking at the sample mean divided by the sample standard deviation

$$T = \frac{\overline{X} - \mu}{S/\sqrt{n}}, \quad \overline{X} = \frac{X_1 + \cdots X_n}{n}, \quad S = \sqrt{\frac{1}{n-1}\sum_{i=1}^{n}(X_i - \overline{X})^2}.$$

This rv will have a t-distribution with $n - 1$ degrees of freedom. Here are the main properties of the t-distribution:

$$ET = 0, \ Var(T) = \frac{k}{k-2}, \ k > 2.$$

2.6.3 F-DISTRIBUTION

This is also a distribution arising in hypothesis testing in statistics as the quotient of two independent χ^2 rvs. In particular,

$$F = \frac{\dfrac{\chi^2(k_1)}{k_1}}{\dfrac{\chi^2(k_2)}{k_2}}.$$

We say $F \sim F(k_1, k_2)$ has an F-distribution with k_1 and k_2 degrees of freedom.

The mean and variance are given by

$$EX = \frac{k_2}{k_2 - 2}, \ k_2 > 2, \ \text{and} \ Var(X) = 2 \left(\frac{k_2}{k_2 - 2} \right)^2 \frac{k_1 + k_2 - 2}{k_1(k_2 - 4)}, \ k_2 > 4.$$

The connection with the t-distribution is

$$F(1, k) = \frac{\chi_1^2(1)/1}{\chi_2^2(k)/k} = \left(\frac{Z}{\sqrt{\chi_2^2(k)/k}} \right)^2, \ Z \sim N(0, 1),$$

which means $F(1, k) = T(k)^2$.

2.7   PROBLEMS

2.1. We roll two dice and X is the difference of the larger number of the two dice and the smaller number. Find $R(X)$, the pmf, and the cdf of X and then find $P(0 < X \le 3)$ and $P(1 \le X < 3)$. Hint: Use the sample space.

2.2. Suppose that the distribution function of X is given by

$$F(b) = \begin{cases} 0 & b < 0 \\ \dfrac{b}{4} & 0 \le b < 1 \\ \dfrac{1}{2} + \dfrac{b-1}{4} & 1 \le b < 2 \\ \dfrac{11}{12} & 2 \le b < 3 \\ 1 & b \ge 3 \end{cases}$$

(a) Find $P(X = i), i = 1, 2, 3$.

(b) Find $P(1/2 < X < 3/2)$.

2.3. If X has cdf $F_X(x)$, what is the cdf

(a) of e^X?

(b) of the random variable $aX + b$, where a and b are nonzero constants?

2.4. Determine c so that $f(x) = P(X = x)$ is a pmf.

(a) $f(x) = \frac{x}{c}, x = 1, 2, 3, \ldots, n$.

(b) $f(x) = \frac{c}{(x+2)(x+3)}, x = 0, 1, 2, 3, \ldots$. Hint: Use partial fractions.

2.5. Let X be a continuous random variable with pdf $f(x) = \begin{cases} 3/4, & 0 \le x \le 1 \\ 1/4, & 2 \le x \le 3 \\ 0, & \text{otherwise.} \end{cases}$

(a) Draw the graph of f.

(b) Determine the cdf F of X, and draw its graph.

2.6. Let $f(x) = \dfrac{1}{\sigma\sqrt{2\pi}} e^{-\dfrac{(x-\mu)^2}{2\sigma^2}}, -\infty < x < \infty$.

(a) Show that f is a pdf.

(b) Show that $x = \mu$ is a max point of f and $x = \mu \pm \sigma$ are inflection points of f. That is, show that $f''(\mu \pm \sigma) = 0$ and $f''(x) < 0, \mu - \sigma < x < \mu + \sigma$ and $f''(x) > 0$ if $-\infty < x < \mu - \sigma$ or $\mu + \sigma < x < \infty$.

2.7. The pdf f of a continuous random variable X is given by:
$$f(x) = \begin{cases} cx + 3, & -3 \le x \le -2 \\ 3 - cx, & 2 \le x \le 3 \\ 0, & \text{otherwise.} \end{cases}$$

(a) Compute c.

(b) Compute the cdf of X.

2.8. The score X of a student on a certain exam is represented by a number between 0 and 1. Suppose that the student passes the exam if this number is at least 0.55. Suppose the pdf of X is given by $f(x) = \begin{cases} 4x, & 0 \le x \le 1/2 \\ 4 - 4x, & 1/2 \le x \le 1 \\ 0, & \text{otherwise.} \end{cases}$

(a) What is the probability that the student fails the exam?

(b) What is the score that he will obtain with a 50% chance, in other words, what is the 50th percentile of the score distribution?

(c) What is the 75th percentile score, i.e., find x_{75} such that $P(X \le x_{75}) = 0.75$?

2.9. Let X be a normal random variable with mean 12 and variance 4. Find the value of c such that $P(X > c) = .10$. This c would be the 90th percentile of X.

2.10. (a) If $X \sim \text{Binom}(n, p)$ show that $P(k \le X \le j) = P(X \le j) - P(X \le k - 1)$. (b) If you toss a fair coin 100 times, what is the probability you get from 52–60 Heads inclusive?

2.11. Suppose 75% of the age group 10–14 years regularly utilize seat belts. Find the probability that in a random stop of 100 automobiles containing 10–14 year olds, 70 or fewer are found to be wearing a seat belt. Find the solution using the binomial distribution as well as the normal approximation to the binomial distribution.

2.12. If the probability that an individual suffers a bad reaction from injection of a given serum is 0.001, determine the probability that out of 2000 individuals (a) exactly 3 and (b) more than 2 individuals will suffer a bad reaction. Calculate this using the binomial, Poisson (with $\lambda = np$), and normal distributions. Which is the better approximation?

2.13. So far in the season, a certain baseball player for the Chicago Cubs has the following probabilities, displayed in the table below, for the outcome of an at-bat. The possible outcomes are an out, a walk, a single, a double, a triple, or a home run.

Outcome of an at-bat	Outcome number	Probability
Out	1	0.662
Walk	2	0.052
Single	3	0.213
Double	4	0.018
Triple	5	0.009
Home run	6	0.046

Find the probability that the player strikes out, walks twice, and doubles twice in the same game.

2.14. For a hypergeometric random variable, determine $\dfrac{P(X = k + 1)}{P(X = k)}$.

2.15. Let $X \sim \text{Binom}(n, p)$ and $Y \sim \text{Poisson}(\lambda = np)$. Compute

(a) $P(X = 2)$ and $P(Y = 2), n = 8, p = 0.1$.

(b) $P(X = 9)$ and $P(Y = 9), n = 10, p = 0.95$.

2.16. If you buy a lotto ticket in 50 games, in each of which your chances of winning is 1/100, what is the probability you will win

(a) at least once?   (b) exactly once?   (c) at least twice?

2.17. Suppose $X = (X_1, \ldots, X_k)$ is multinomial with parameters n, k and $p_1, \ldots, p_k$. Show that $P(X_i = x) = \displaystyle\sum_{x_j, j \neq i} P(X_1 = x_1, \ldots X_i = x, \ldots, X_k = x_k) = \binom{n}{x} p_i^x (1 - p_i)^{n-x}$
so that each $X_i \sim \text{Binom}(n, p_i)$. Hint: Define success as in category i, failure as not in category i.

2.18. If X is Geometric(p), show that $P(X \geq n + k | X \geq n) = P(X \geq k)$. This is the memoryless property since it implies that a geometric rv does not recall that n trials have already passed.

2.19. Let X be a random variable that takes values in $[0, 1]$, and has cdf given by $F_X(x) = x^2, 0 \leq x \leq 1$, with $F_X(x) = 0, x < 0$, $F_X(x) = 1, x > 1$. Compute $P(1/2 < X \leq 3/4)$ and find the pdf of X.

2.20. Suppose we choose arbitrarily a point from the square with corners at (2,1), (3,1), (2,2), and (3,2). The random variable A is the area of the triangle with its corners at (2,1), (3,1) and the chosen point.

(a) What is the largest area that can occur, and what is the set of points for which $A \leq 1/4$?

(b) Determine the distribution function F of A.

(c) Determine the pdf f of A.

2.21. Show that if Z is a standard normal random variable, then, for $x > 0$,

(a) $P(Z > x) = P(Z < -x)$;

(b) $P(|Z| > x) = 2P(Z > x)$; and

(c) $P(|Z| < x) = 2P(Z < x) - 1.$

2.22. Jensen, arriving at a bus stop, just misses the bus. Suppose that he decides to walk if the (next) bus takes longer than 5 minutes to arrive. Suppose also that the time in minutes between the arrivals of buses at the bus stop is a continuous random variable with a Unif(4, 6) distribution. Let X be the time that Jensen will wait.

    (a) What is the probability that X is less than 4 1/2 (minutes)?

    (b) What is the probability that X equals 5 (minutes)?

    (c) Is X a discrete random variable or a continuous random variable?

2.23. Let X have an Exp(0.2) distribution. Compute $P(X > 5)$.

2.24. Let $X \sim \text{Exp}(\lambda), \lambda > 0$. Find a value m such that $P(X \leq m) = 0.5$.

2.25. Let $Z \sim N(0, 1)$. Find a number z such that $P(Z \leq z) = 0.9$. Also find z so that $P(-z \leq Z \leq z) = 0.9$.

2.26. The time (in hours) required to repair a machine is an exponentially distributed random variable with parameter $\lambda = 1.2$

    (a) What is the probability that a repair time exceeds 2 hours?

    (b) What is the conditional probability that a repair takes at least 10 hours, given that its duration exceeds 9 hours?

2.27. The number of years a radio functions is exponentially distributed with parameter $\lambda = 1/18$. If Jones buys a used radio, what is the probability that it will work for an additional 8 years?

2.28. A patient has insurance that pays $1,500 per day up to 3 days and $1,000 per day after 3 days. For typical illnesses the number of days in the hospital, X, has the pmf $p(k) = \frac{7-k}{21}, k = 1, 2, \ldots, 7$. Find the mean expected amount the insurance company will pay.

2.29. Let $X \sim N(\mu, \sigma)$ use substitution and integration by parts to verify $E(X) = \mu$ and $SD(X) = \sigma$. That is, verify Example 2.36.

2.30. An investor has the option of investing in one of two stocks. If he buys Stock A he can net $500 with probability 1/2, and lose $100 with probability 1/2. If he buys Stock B, he can net $1,500 with probability 1/4 and lose $200 with probability 3/4.

    (a) Find the mean and SD for each investment. Which stock should he buy based on the **coefficient of variation** defined by SD/μ?

    (b) What is the interpretation of the coefficient of variation?

(c) The value of x dollars is worth $g(x) = \sqrt{x + 200}$ to the investor. This is called a utility function. What is the expected utility to the investor for each stock?

2.31. Suppose an rv has density $f(x) = 2x, 0 < x < 1$ and 0 otherwise. Find

(a) $P(X < 1/2), P(1/4 < X \leq 1/2), P(X < 3/4 | X > 1/2)$ and

(b) $E[X], SD[X], E[e^{tX}]$.

2.32. An rv has pdf $f(x) = 5e^{-5x}, 0 \leq x < \infty$, and 0 otherwise. Find $E[X], Var[X], med[X]$.

2.33. Find the mgf of $X \sim$ Geometric(p). (Hint: $\sum_{k=1}^{\infty} a^k = \frac{a}{1-a}, |a| < 1$.) Use it to find the mean and variance of X.

2.34. Find the mgf of $X \sim$ Poisson(λ) (Hint: $\sum_{k=0}^{\infty} \frac{a^k}{k!} = e^a$.) Use it to find the mean and variance of X.

2.35. Suppose X has the mgf $M_X(t) = 0.09e^{-2t} + 0.24e^{-t} + 0.24e^{t} + 0.09e^{2t} + 0.34$.

(a) Find $P(X \leq 0)$.

(b) Find $E[X]$.

2.36. Let $X \sim$ Unif$[0, 1]$. Find $E[4X^5 + 5X^4 + 4X^3 - 8X^2 + 7X + 1]$.

2.37. The mean deviation of a discrete rv is defined by $MD(X) = \sum_{\{x \,|\, P(X=x)>0\}} |x - E[X]| P(X = x)$. Find the mean deviation of the rv X which is the sum of the faces of two fair dice which are rolled.

2.38. The mean deviation of a continuous rv X with pdf $f_X(x)$ is defined by $MD(X) = \int_{-\infty}^{\infty} |x - E[X]| f_X(x)\, dx$. Find the mean deviation for $X \sim$ Exp(λ) and for $X \sim$ Unif$[a, b]$.

2.39. An exam is graded on a scale 0–100. A student must score at least 60 to pass. Student scores are modeled by the density $f(x) = \begin{cases} 0.0004x & 0 \leq x \leq 50 \\ 0.04 - 0.0004x & 50 \leq x \leq 100 \\ 0 & \text{otherwise.} \end{cases}$

(a) Find the probability a student passes.

(b) What exam score is the 85th percentile?

2.40. Find EX and $Var(X)$ for the rv X with the following densities.

(a) $P(X = k) = \frac{1}{n}, k = 1, 2, \ldots, n$.

(b) $f_X(x) = rx^{r-1}, 0 < x < 1, r > 0$.

2.41. Let N be the number of Bernoulli trials (meaning independent and only one of two outcomes possible) to get r successes. N is a negative binomial rv with parameters r, p, NegBinom(r, p), and we know $P(N = k) = \binom{k-1}{r-1} p^r (1 - p)^{k-r}$, $k = r, r + 1, r + 2, \ldots$. If you think of the process restarting after each success is obtained, it is reasonable to write $N = Y_1 + \cdots + Y_r$, where Y_i's are independent geometric rvs. Use this to find the mgf of N and then find EN, $Var(N)$.

2.42. Let X be Hypergeometric(N, n, k). Let $p = k/N$. It can be shown that $EX = np$ and $Var(X) = np(1 - p)\frac{N-n}{N-1}$. This looks like the same mean and variance of a Binomial(n, p) with $p = \dfrac{k}{N}$ except for the extra term $\dfrac{N - n}{N - 1}$. This term is known as the **Correction factor**. We know that Binomial(n, p) can be approximated by Normal$(np, \sqrt{np(1 - p)})$. What is the approximate distribution of Hypergeometric(N, n, k)?

2.43. We throw a coin until a head turns up for the second time, where p is the probability that a throw results in a head and we assume that the outcome of each throw is independent of the previous outcomes. Let X be the number of times we have to throw the coin.

    (a) Determine $P(X = 2)$, $P(X = 3)$, and $P(X = 4)$.

    (b) Show that $P(X = n) = (n - 1)p^2(1 - p)^{n-2}$, for $n \geq 2$.

    (c) Find EX.

2.44. Suppose $P(X = 0) = 1 - P(X = 1)$, $E(X) = 3Var(X)$. Find $P(X = 0)$.

2.45. If $EX = 1$, $Var(X) = 5$ find $E(2 + X)^2$ and $Var(4 + 3X)$.

2.46. Monthly worldwide major mainframe sales is 3.5 per year and has a Poisson distribution. Find

    (a) the probability of at least 2 sales in the next month,

    (b) the probability at most one sale in the next month, and

    (c) the variance of the monthly number of sales.

2.47. A batch of 100 items has 6 defective and 94 good. If X is the number of defectives in a randomly drawn sample of 10, find $P(X = 0)$, $P(X > 2)$, EX, $Var(X)$.

2.48. An insurance company sells a policy with a 1 unit deductible. Let X be the amount of the loss have pmf $f(x) = \begin{cases} 0.9, x = 0 \\ c/x, x = 1, 2, 3, 4, 5, 6. \end{cases}$

Find c and the expected total amount the insurance company has to pay out.

2.49. Find EX, $E[X(X - 1)]$, if X has pmf $f(x) = \binom{4}{x}\left(\frac{1}{2}\right)^4$, $x = 0, 1, 2, 3, 4$.

2.50. Choose a so as to minimize $E[|X - a|]$:

(a) when X is uniformly distributed over $(0, A)$ and

(b) when X is now exponential with rate λ.

2.51. Suppose that X is a normal random variable with mean 5. If $P(X > 9) = .2$, what is $Var(X)$?

2.52. We have the rvs X, Y with the joint density $f(x, y) = x + y$ for $0 \leq x, y \leq 1$, and $f(x, y) = 0$ otherwise. Find the marginal densities $f_X(x)$ and $f_Y(y)$. Are X and Y independent? Calculate $E(X + Y)$.

2.53. Suppose a population contains 16% ex-cons. If 50 people are selected at random use the CLT to estimate $P(X < 5)$ where X is the number of ex-cons in your sample.

2.54. Show that (a) $Cov(X, Y) = Cov(Y, X)$, and (b) $Cov(aX + b, Y) = aCov(X, Y)$.

2.55. Let $c = k\sigma$ in Chebychev's inequality. Show that $P(|X - \mu| < k\sigma) \geq 1 - \frac{1}{k^2}$. Estimate the probability that the values of a rv fall within two standard deviations of the mean.

2.56. Consider the discrete rv with pdf $P(X = \pm 2) = \frac{1}{8}$, $P(X = 0) = \frac{3}{4}$. Find EX, $Var(X)$ and use Chebychev to find $P(|X| \geq 2)$ and compare it with the exact answer.

2.57. Let $X \sim Poisson(\Lambda)$ where $\Lambda \sim Unif[0, 3]$. Find $P(X = 1)$ by using the Law of Total Probability $P(X = 1) = \int_0^3 P(X = 1 | \Lambda = \lambda) P(\Lambda = \lambda)\, d\lambda$.

CHAPTER 3

Distributions of Sample Mean and Sample SD

This chapter begins the probabilistic study of statistics. When we take a random sample from the population we wind up with new random variables like the Sample Mean $\overline{X}$ and Sample Standard Deviation S and various functions involving these quantities. In statistics we need to know the probabilistic distribution of these quantities in order to be able to quantify the errors we make in approximating parameters like the population mean using the random sample.

3.1   POPULATION DISTRIBUTION KNOWN

A random variable representing the item of interest is labeled X and is called the **population** random variable. This could be something like the income level of an individual or the voter preference or the efficacy of a drug, etc. A **random sample from** X is a collection of **independent random variables** $X_1, X_2, \ldots, X_n$, **which have the same distribution as** X. The random variables in the sample have the same pdf and cdf as the population random variable X. Because the random variables $X_1, X_2, \ldots, X_n$, are independent and with the same distribution, this is a model of sampling **with replacement**.

A population box is a representation of each individual in the population using a ticket. On the ticket is written the particular object of interest for study, such as weight, income, IQ, years of schooling, etc.

The box contains one ticket for each individual in the population and the number on the ticket is the item of interest to the experimenter. Think of a random sample as choosing tickets from the population box with replacement (so the box remains the same on each draw). If we don't replace the tickets, the box changes after each draw and the random variables would no longer be independent or have the same distribution as X. Each time we take a sample from the population we are getting values $X_1 = x_1, X_2 = x_2, \ldots, X_n = x_n$ and these values $(x_1, x_2, \ldots, x_n)$ are specific **observed values** of the random variables $(X_1, \ldots, X_n)$. In general, lowercase variables will be observed variables, while uppercase represents random variables before observation.

Once we have a random sample we want to summarize the values of the random variables to obtain some information.

Definition 3.1   Given any collection of random variables $X_1, X_2, \ldots, X_n$, the **sample mean** is $\overline{X} = \frac{X_1+X_2+\cdots+X_n}{n}$. The **sample variance** is $S^2 = \frac{1}{n-1} \sum_{i=1}^{n}(X_i - \overline{X})^2$ and the **sample standard deviation** is S. The **sample median**, assuming the random variables are sorted, is

$$\widetilde{X} = \begin{cases} X_{\frac{n+1}{2}}, & \text{if } n \text{ is odd;} \\ \frac{X_{\frac{n}{2}}+X_{\frac{n}{2}+1}}{2}, & \text{if } n \text{ is even.} \end{cases}$$

Any function $g(X_1, \ldots, X_n)$ of the random sample is said to be a **statistic**. For example $\overline{X}$, S, and $\widetilde{X}$ are examples of statistics.

Remark 3.2   It is important to keep in mind that $\overline{X}, S$, and $\widetilde{X}$ are random variables, not numbers. Once the experiment is performed and we have observations $X_1 = x_1, \ldots, X_n = x_n$, then the **observed statistics**

$$\overline{x} = \frac{x_1 + x_2 + \cdots + x_n}{n}, s^2 = \frac{1}{n-1} \sum_{i=1}^{n}(x_i - \overline{x})^2, \widetilde{x} = \begin{cases} x_{\frac{n+1}{2}}, & \text{if } n \text{ is odd;} \\ \frac{x_{\frac{n}{2}}+x_{\frac{n}{2}+1}}{2}, & \text{if } n \text{ is even} \end{cases}$$

result in real numbers and these are not random variables.

Example 3.3   Consider the population box $\boxed{0}\,\boxed{1}\,\boxed{2}\,\boxed{3}\,\boxed{4}$. This is another way of saying the population rv is discrete with $R(X) = \{0, 1, 2, 3, 4\}$ and $P(X = k) = 1/5, k = 0, 1, 2, 3, 4$. We will choose random samples of size 2 from this population, **without replacement**. The **population mean** of the numbers on the tickets is $\mu = E(X) = \frac{0+1+2+3+4}{5} = 2$ with **population variance**$= E(X - 2)^2 = \frac{1}{5} \sum_{i=1}^{5}(x_i - 2)^2 = 2$.

How many samples of size 2 are there? Since we don't care about the order, we are asking for how many combinations of the 5 numbers taken 2 at a time there are and that is $\binom{5}{2} = 10$. Here are the 10 possible samples of size 2:

$$(0, 1), (0, 2), (0, 3), (0, 4), (1, 2), (1, 3), (1, 4), (2, 3), (2, 4), (3, 4).$$

We take the sample mean $\overline{X} = \frac{X_1+X_2}{2}$, where X_i is the number on ticket $i = 1, 2$. The possible values of $\overline{X}$ are $1/2, 1, 3/2, 2, 5/2, 3, 7/2$. The distribution of $\overline{X}$ is

x	1/2	1	3/2	2	5/2	3	7/2
$P(\overline{X} = x)$	0.1	0.1	0.2	0.2	0.2	0.1	0.1

For example, $P(\overline{X} = 3/2) = 0.2$ because there are 10 samples of size 2 of which exactly 2 have a sample mean of $3/2$. Now that we have the distribution of $\overline{X}$, we may compute

$$E[\overline{X}] = \sum_{i=1}^{7} \overline{x}_i P(\overline{X} = \overline{x}_i)$$

$$= 0.1(1/2) + 0.1(1) + 0.2(3/2) + 0.2(2) + 0.2(5/2) + 0.1(3) + 0.1(7/2) = 2$$

the same as the population mean μ. This is not a coincidence. Next, the variance of the sample mean is

$$Var(\overline{X}) = \sum_{i=1}^{7} (\overline{x}_i - 2)^2 P(\overline{X} = \overline{x}_i) = 0.75.$$

The population variance is 2 but the variance of the sample mean is 0.75. The variance of the sample mean is always lower than the variance of the individuals in the population as we will show in the next theorem. Averages have lower variation.

The theorem will show us how to calculate $E(\overline{X})$ and $SD(\overline{X})$ directly from the population mean, SD, and sample size.

Theorem 3.4 If $X_1, \ldots, X_n$, is a random sample from a population X with mean $\mu = EX$ and variance $\sigma^2 = Var(X)$, then

$$\boxed{E[\overline{X}] = \mu, \qquad Var(\overline{X}) = \frac{\sigma^2}{n}.}$$

If the **population size, N, is finite**, and the sampling is done **without replacement**, then

$$\boxed{E[\overline{X}] = \mu \qquad Var(\overline{X}) = \frac{\sigma^2}{n} \frac{N-n}{N-1}.}$$

In our example, we do have a finite population $N = 5$ so the variance of $\overline{X}$ calculated using the theorem is $Var(\overline{X}) = \frac{\sigma^2}{n} \frac{N-n}{N-1} = \frac{2}{2} \frac{5-2}{5-1} = \frac{3}{4}.$

To show the theorem, we have

$$E[\overline{X}] = \frac{1}{n} E[X_1 + \cdots + X_n] = \frac{1}{n} n\mu = \mu.$$

Also, we know

$$Var(X + Y) = Var(X) + 2Cov(X, Y) + Var(Y).$$

If X, Y are uncorrelated, which is certainly true if they are independent, then $Var(X + Y) = Var(X) + Var(Y)$. Therefore,

$$Var(\overline{X}) = \frac{1}{n^2}[Var(X_1) + \cdots + Var(X_n)] = \frac{n\sigma^2}{n^2} = \frac{\sigma^2}{n}.$$

The rest of the proof when sampling is done without replacement is based on the Hypergeometric distribution and is omitted. □

Definition 3.5   The **Standard Error** (*SE*) of the sample mean is $SE = SD(\overline{X}) = \sqrt{E(\overline{X} - E(\overline{X}))^2}$. When we are sampling from an infinite population or from a finite population but with replacement, then $SE(\overline{X}) = \frac{\sigma}{\sqrt{n}}$. When we are sampling from a **finite population without replacement**, then

$$SE(\overline{X}) = \sqrt{\frac{N-n}{N-1}} \times \frac{\sigma}{\sqrt{n}}.$$

The term $\sqrt{\frac{N-n}{N-1}}$, when we have a finite population and sampling is without replacement, is called the **finite population correction factor for the standard error of the sample mean**.

Remark 3.6   Theorem 3.4 shows that $E\overline{X} = \mu$. Whenever we have an estimator, in this case $\overline{X}$, of a parameter inherent to the population rv, in this case μ, and we know $E\overline{X} = \mu$, we say that the statistic $\overline{X}$ is an **unbiased estimator** of μ.

Sample means are calculated using $\overline{X} = (X_1 + \cdots + X_n)/n$. Sometimes we are interested in the distribution of the sums $S_n = X_1 + \cdots X_n$. Here is the result.

Theorem 3.7   If $X_1, \ldots, X_n$ is a random sample from a population X with mean $\mu = EX$ and variance $\sigma^2 = Var(X)$, then

$$E\left[\sum_{i=1}^{n} X_i\right] = n\mu, \qquad Var\left[\sum_{i=1}^{n} X_i\right] = n\sigma^2.$$

Thus, $SE(S_n) = \sigma\sqrt{n}$. If we have a **finite population and we sample without replacement**, $E[S_n] = n\mu$ but $SE(S_n) = \sigma\sqrt{n}\sqrt{\frac{N-n}{N-1}}$.

Since $nS_n = \overline{X} \implies E[S_n] = nE[\overline{X}] = n\mu$, and $Var[S_n] = Var[n\overline{X}] = n^2 Var[\overline{X}] = n^2\frac{\sigma^2}{n} = n\sigma^2$. The point is that we just multiply the result for $\overline{X}$ by n.

3.1.1 THE POPULATION $X \sim N(\mu, \sigma)$

In this section we assume the population rv X actually has a Normal distribution with unknown mean μ but known standard deviation σ. The goal is to estimate the unknown parameter μ by using the sample mean $\overline{X}$.

Theorem 3.8 If $X \sim N(\mu, \sigma)$, and $X_1, \ldots, X_n$ is a random sample from X, then $\overline{X} \sim N(\mu, \frac{\sigma}{\sqrt{n}})$.

Proof. In Proposition 2.61 we see that $S_n = X_1 + \cdots + X_n \sim N(n\, \mu, \sqrt{n}\, \sigma)$. Therefore, $\overline{X} = \frac{S_n}{n} \sim N(\mu, \frac{\sigma}{\sqrt{n}})$. $\qquad\square$

Now that we know $\overline{X} \sim N(\mu, \frac{\sigma}{\sqrt{n}})$ we may standardize $\overline{X}$ to see that $Z = \frac{\overline{X} - \mu}{\sigma/\sqrt{n}}$ is $N(0, 1)$. We can answer questions about the chances the sample mean lies in a particular interval $P(a \leq \overline{X} \leq b)$. Here's an example

Example 3.9 Suppose IQs of a population are normally distributed $N(100, 10)$. We know $\overline{X} \sim N(100, \frac{10}{\sqrt{n}})$. If we take a random sample of $n = 25$ people, we find

(a) $P(95 < \overline{X} < 105) = \text{normalcdf}(95, 105, 100, 10/\sqrt{25}) = 0.9875$.

(b) $P(\overline{X} > 120) = \text{normalcdf}(120, \infty, 100, 10/\sqrt{25}) = 7.7 \times 10^{-24} \approx 0$.

(c) $P(\overline{X} < 98) = \text{normalcdf}(-\infty, 98, 100, 10/\sqrt{25}) = 0.1586$.

You can see that the sample mean $\overline{X}$ has much less variability than X itself since, for example $P(95 < X < 105) = \text{normalcdf}(95, 105, 100, 10) = 0.3829$. This says about 38% of individuals have IQs between 95 and 105, but 98.75% of samples of size 25 will result in a sample mean between 95 and 105. Individuals may vary a lot, but averages don't.

Many underlying populations do not follow a normal distribution. For instance, the distributions of annual incomes, or times between arrivals of customers, are not normally distributed. Now we ask the question what happens to the sample mean when the underlying population does not follow a normal distribution.

3.1.2 THE POPULATION X IS NOT NORMAL BUT HAS KNOWN MEAN AND VARIANCE

Here's where the Central Limit Theorem 2.63 plays a big role. Even though X, having mean $EX = \mu$ and $SD(X) = \sigma$, is not normal, the Theorem 2.63 says that $\overline{X}$ will be approximately $N(\mu, \frac{\sigma}{\sqrt{n}})$ as long as the sample size n is large enough. Specifically, the CLT theorem says that for $Z \sim N(0, 1)$,

$$\lim_{n \to \infty} P\left(a \le \frac{\overline{X} - \mu}{\sigma/\sqrt{n}} \le b\right) = P\left(a \le Z \le b\right). \tag{3.1}$$

We will use this in two forms, one for the sample mean $\overline{X}$ and the other for the sum $S_n = X_1 + \cdots + X_n$.

(a) $\overline{X} \approx N(\mu, \frac{\sigma}{\sqrt{n}})$.

(b) $S_n = X_1 + X_2 + \cdots + X_n \approx N(n\mu, \sqrt{n}\sigma)$.

How large does n need to be to make the approximation decent? An exact answer depends on the underlying population distribution but a good rule of thumb is $n \ge 30$.

Example 3.10   Suppose we take a random sample of size $n = 10$ from a population $X \sim$ Unif[0, 10]. We want to find $P(\overline{X} \le 4)$. Even though $n < 30$ let's use the normal approximation. We know $\overline{X} \approx N(\mu, \frac{\sigma}{\sqrt{n}})$ and in this case $\mu = EX = 5$ and $Var(X) = \sigma^2 = 10^2/12$. Therefore, $\overline{X} \approx N(5, 10/\sqrt{12 \cdot 10}) = N(5, 0.9128)$. Using this we get $P(\overline{X} \le 4) \approx$ normalcdf$(0, 4, 5, 0.9128) = 0.1366$.

Analytically it is not easy to find $P(\overline{X} \le 4) = P(X_1 + \cdots + X_{10} \le 40)$ directly using the uniform distribution because the sum of 10 independent uniform rvs is not uniform. Nevertheless, we may use technology to find the exact value of $P(X_1 + \cdots + X_{10} \le 40) = 0.138902$. Even with $n = 10$ we get a really good approximation of 0.1366. In Figure 3.1 we show what happens to the pdf of the sum of n Uniform[0,10] distributions if we take $n = 1, 2, 4, 10$. You can see that with $n = 10$ the pdf looks a lot like a normal density. Even with $n = 4$ it looks normal (see Figure 3.1).

3.1.3   THE POPULATION IS BERNOULLI, p KNOWN

There are many problems in statistics in which each individual in the population is either a success (represented by a 1) or a failure (represented by a 0). The box model is

$$\boxed{0}\,\boxed{1}\,\boxed{1}\,\boxed{0}\ \ldots \quad p = \text{proportion of 1s in the box.}$$

Suppose we know the probability of success is $p, 0 < p < 1$. This also means the percentage of 1s in the population will be $100p\%$. The population rv is $X \sim$ Bernoulli(p) and we take a random sample $X_1, \ldots, X_n$ from X. Each X_i is either 0 or 1 and $S_n = X_1 + \cdots + X_n$ is the total number of 1s in our sample and $\overline{X} = \frac{S_n}{n}$ is the fraction of 1s in the sample. It is also true that S_n is exactly Binom(n, p). By Theorem 2.63 we also know

$$\boxed{S_n \approx N(np, \sqrt{np(1-p)}) \text{ and } \overline{X} \approx N\left(p, \sqrt{\frac{p(1-p)}{n}}\right).}$$

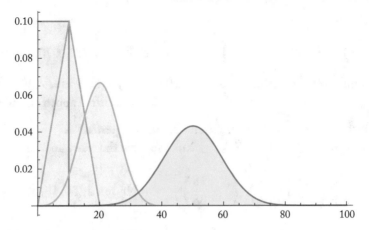

Figure 3.1: Sum of $n = 1, 2, 4, 10$ Uniform$[0, 10]$ distributions.

In general, we will use this approximation as long as $np \geq 5$ and $n(1 - p) \geq 5$.

Example 3.11   Suppose a gambler bets \$1 on Red in a game of roulette. The chance he wins is $p = \frac{18}{38}$ and the bet pays even money (which means if red comes up he is paid his \$1 plus an additional \$1). He plans to play 50 times.

First, how many games do we expect to win? That is given by $E S_{50} = 50 \times \frac{18}{38} = 23.68$ and we will expect to lose $50 - 23.68 = 26.32$ games. Since the bet pays even money and we are betting \$1 on each game, we expect to lose \$26.32.

How far off this number do we expect to be, i.e., what is the SE for the number of games won and lost? The SE is $SD(S_{50}) = \sqrt{50\frac{18}{38}(1 - \frac{18}{38})} = 3.53$ so we expect to win 23.68 games, give or take about 3.53 games. Put another way, we expect to lose \$26.32, give or take \$3.53.

(a) What is the chance he wins 40% of the games, i.e., 20 games?
We are looking for $P(\overline{X} = 0.4)$ which is equivalent to $P(S_{50} = 20)$. We'll do this in two ways to get the exact result and then using the normal approximation.

First, since $S_{50} \sim \text{Binom}(50 \times \frac{18}{38}, \sqrt{50\frac{18}{38}(1 - \frac{18}{38})} = \text{Binom}(23.68, 3.53)$ we get the exact value

$$P(S_{50} = 20) = \binom{50}{20}\left(\frac{18}{38}\right)^{20}\left(1 - \frac{18}{38}\right)^{30} = \text{binompdf}(50, 18/38, 20) = 0.066.$$

Using the normal approximation we have to use the continuity correction because the probability a continuous rv is any single value is always zero. We use $\overline{X} \approx N\left(\frac{18}{38}, \sqrt{\frac{18/38(1-18/38)}{50}}\right) =$

$N(0.473, 0.0706)$

$$P(\overline{X} = 0.4) \approx P(.39 \leq \overline{X} \leq .41) = \text{normalcdf}(.39, .41, .473, .0706) = 0.06623.$$

Note that the continuity correction would use $19.5 \leq S_{50} \leq 20.5 \implies .39 \leq \overline{X} \leq .41$. Also, $np = 50 \times 18/38 > 5$ and $n(1 - p) = 50 \times 20/38 > 5$, so a normal approximation may be used.

(b) What is the chance he wins no more than 40% of the games?
   Now we want $P(\overline{X} \leq 0.4)$ or $P(S_{50} \leq 20)$. We have the exact answer

$$P(S_{50} \leq 20) = \sum_{k=0}^{20} \binom{50}{k} \left(\frac{18}{38}\right)^k \left(1 - \frac{18}{38}\right)^{50-k} = \text{binomcdf}(50, 18/38, 20) = 0.1837.$$

The approximate answer is either given by $P(S_{50} \leq 20) \approx \text{normalcdf}(0, 20.5, 23.68, 3.53) = 0.18383$ or $P(\overline{X} \leq 0.4) = \text{normalcdf}(0, 0.4, .473, .0706) = .1505$. If we use the continuity correction, it will be $P(\overline{X} \leq 0.4) = \text{normalcdf}(0, 0.41, .473, .0706) = .18610$.

(c) What is the chance the gambler comes out ahead?
   This is asking for $P(S_{50} \geq 25)$ or $P(\overline{X} \geq 0.5)$ Using the same procedure as before we get the exact answer

$$P(S_{50} \geq 25) = 1 - P(S_{50} \leq 24) = 1 - \text{binomcdf}((50, 18/38, 24) = .4078.$$

Then, since $24.5/50 = .49$ with the continuity correction we get

$$P(\overline{X} \geq .49) \approx \text{normalcdf}(.49, 1, .473, .0706) = 0.40485.$$

Without the continuity correction $P(\overline{X} \geq .5) = .35106$.

   The result for Bernoulli populations, i.e., when X is either 0 or 1, extends to a more general case when we assume the population takes on any two values. Here is the result.

Theorem 3.12   Suppose the population is $X = \begin{cases} a, & \text{with probability } p, \\ b, & \text{with probability } 1 - p, \end{cases}$ and we have a random sample $X_1, \ldots, X_n$, from the population. Then

   (a) $EX = ap + b(1 - p)$ and $Var(X) = (a - b)^2 p(1 - p)$.

   (b) $E S_n = E(X_1 + \cdots + X_n) = n(ap + b(1 - p))$ and $Var(S_n) = n(a - b)^2 p(1 - p)$.

   (c) $E\overline{X} = ap + b(1 - p)$ and $Var(\overline{X}) = \frac{1}{n}(a - b)^2 p(1 - p)$.

   (d) $S_n \approx N(n(ap + b(1 - p)), |a - b| \sqrt{np(1 - p)})$
       and $\overline{X} \approx N(ap + b(1 - p), |a - b| \sqrt{\frac{p(1-p)}{n}})$.

Example 3.13   Going back to Example 3.11 we can think of each play with outcome the amount won or lost. That means the population rv is

$$X = \begin{cases} +1, & \text{with probability } 18/38; \\ -1, & \text{with probability } 20/38. \end{cases}$$

This results in $EX = -0.0526, SD(X) = 0.9986$. Therefore, the mean expected winnings in 50 plays is $ES_{50} = -2.63$ with SE, $SD(S_{50}) = 7.06$. In 50 plays we expect to lose -2.63 dollars, give or take 7.06. Now we ask what are the chances of losing more than \$4? This is $P(S_{50} < -4) \approx \text{normalcdf}(-\infty, -4, -2.63, 7.06) = 0.423$. Using the continuity correction $P(S_{50} < -4) \approx \text{normalcdf}(-\infty, -4.5, -2.63, 7.06) = 0.395$.

Example 3.14   A multiple choice exam has 20 questions with 4 possible answers for each question. A student will guess the answer for each question by choosing an answer at random. To penalize guessing, the test is scored as $+2$ for each correct answer and -1 for each incorrect answer.

(a) What is the students's expected score and the SE for the score?
Answering a question is just like choosing at random 1 out of 4 possible tickets from the box $\boxed{-1}\,\boxed{-1}\,\boxed{-1}\,\boxed{2}$. Since there are 20 questions, we will do this 20 times and total up the numbers on the tickets. The population mean is $(2)\frac{1}{4} + (-1)\frac{3}{4} = -\frac{1}{4}$ and the population SD is $(2 - (-1))\sqrt{\frac{1}{4}\frac{3}{4}} = 1.299$.
The expected score is $ES_{20} = 20(-\frac{1}{4}) = -5$ with $SE = SD(S_{20})\sqrt{20} \times 1.299 = 5.809$. The best estimate for the student's score is -5, give or take about 6.

(b) Find the approximate chance the student scores a 5 or greater.
Since $S_{20} \approx N(-5, 5.809)$ we have $P(S_{20} \geq 4.5) = 0.0509$.

3.2   POPULATION VARIANCE UNKNOWN: SAMPLING DISTRIBUTION OF THE SAMPLE VARIANCE

Up until now we have assumed that the population mean and variance were known. When μ is unknown we are going to use $\overline{X}$ to approximate it. In this section we show what happens when σ is unknown and use the sample SD to approximate it.

We have a random sample $X_1, \ldots, X_n$ from a population X assumed to have $EX = \mu$ and $Var(X) = \sigma^2$. Recall that the sample variance is given by

$$S^2 = \frac{1}{n-1} \sum_{i=1}^{n} (X_i - \overline{X})^2.$$

If we use S^2 to approximate σ^2 can we expect it to be a good approximation? And why do we divide by $n-1$ instead of just n? The next theorem answers both of these questions.

Theorem 3.15 S^2 is an unbiased estimator of σ^2. That is, $ES^2 = \sigma^2$

Proof. If $\overline{X}$ was replaced by μ, we could make this easy calculation

$$E\left(\frac{1}{n-1}\sum_{i=1}^{n}(X_i - \mu)^2\right) = \frac{1}{n-1}\sum_{i=1}^{n}E(X_i - \mu)^2 = \frac{n}{n-1}\sigma^2.$$

But we have the factor $\frac{n}{n-1}$ which shouldn't appear, so it must come from the substitution of $\overline{X}$ for μ. Here is what we have to do:

$$E\left(\frac{1}{n-1}\sum_{i=1}^{n}(X_i - \overline{X})^2\right) = E\left(\frac{1}{n-1}\sum_{i=1}^{n}(X_i - \mu + \mu - \overline{X})^2\right)$$

$$= \frac{1}{n-1}E\left(\sum_{i=1}^{n}(X_i - \mu)^2 + 2(X_i - \mu)\left(\mu - \overline{X}\right) + \left(\mu - \overline{X}\right)^2\right)$$

$$= \frac{1}{n-1}\left(n\sigma^2 + nE\left(\mu - \overline{X}\right)^2 + 2E\left(\mu - \overline{X}\right)(X_i - \mu)\right)$$

$$= \frac{1}{n-1}\left(n\sigma^2 + n\frac{\sigma^2}{n} + 2E(\mu - \overline{X})(n\overline{X} - n\mu)\right) = \frac{1}{n-1}\left(n\sigma^2 + \sigma^2 - 2nE(\overline{X} - \mu)^2\right)$$

$$= \frac{1}{n-1}\left(n\sigma^2 + \sigma^2 - 2n\frac{\sigma^2}{n}\right) = \frac{\sigma^2}{n-1}(n+1-2) = \sigma^2.$$

That's why we divide by $n-1$ instead of just n. If we divided by n, S^2 would be a biased estimator of σ^2.   □

Now we need to determine not just ES^2 but the distribution of S^2.

Theorem 3.16   If we have a random sample from a **normal population** $X \sim N(\mu, \sigma)$, then $\frac{n-1}{\sigma^2}S^2 \sim \chi^2(n-1)$. In particular, $ES^2 = \frac{\sigma^2}{n-1}E\chi^2(n-1) = \sigma^2$ and

$$Var(S^2) = Var\left(\frac{\sigma^2}{n-1}\chi^2(n-1)\right)$$

$$= \frac{\sigma^4}{(n-1)^2}Var(\chi^2(n-1)) = \frac{\sigma^4}{(n-1)^2}2(n-1) = 2\frac{\sigma^4}{n-1}.$$

We will skip the proof of this theorem but just note that if we replaced $\overline{X}$ by μ, we would have

$$W^2 = \frac{1}{n-1}\sum_{i=1}^{n}(X_i - \mu)^2 = \frac{\sigma^2}{n-1}\sum_{i=1}^{n}\left(\frac{X_i - \mu}{\sigma}\right)^2 = \frac{\sigma^2}{n-1}\sum_{i=1}^{n}Z_i^2,$$

where $Z_i \sim N(0,1)$, $i = 1,2,\ldots,n$ is a set of n independent standard normal rvs. But we know that in this case $\sum_{i=1}^{n}Z_i^2 \sim \chi^2(n)$. Consequently, $\frac{n-1}{\sigma^2}W^2 \sim \chi^2(n)$. That looks really close to what the theorem claims and we would be done except for the fact that $W \neq S$ and $\chi^2(n) \neq \chi^2(n-1)$. Replacing μ by $\overline{X}$ accounts for the difference. We omit the details.

Example 3.17   Suppose we have a population $X \sim N(\mu, 10)$ and we choose a random sample $X_1,\ldots, X_{25}$ from X. Letting S^2 denote the sample variance, we want to find the following.

(a) $P(S^2 > 50)$. Using the theorem we have $(n-1)S^2/\sigma^2 \sim \chi^2(24)$ so that $P(S^2 > 50) = P(24S^2/100 > 24(50)/100) = P(\chi^2(24) > 12) = \chi^2 cdf(12, \infty, 24) = 0.9799$.

(b) $P(75 < S^2 < 125) = P(.24(75) < \chi^2(24) < .24(125)) = 0.61825$.

(c) $E S^2 = \sigma^2 = 100$, $Var(S^2) = \dfrac{2\sigma^4}{n-1} = 10{,}000/12$.

Now we know that if σ was known and we had a random sample from a normal population $X \sim N(\mu, \sigma)$, then $\frac{\overline{X}-\mu}{\sigma/\sqrt{n}} \sim N(0,1)$. If σ is unknown, this is no longer true. What we want is to replace σ by S, which is determined from the sample and not the population. That makes the denominator also depend on the random sample. Consider the rv

$$T = \frac{\overline{X}-\mu}{\frac{S}{\sqrt{n}}} = \frac{\frac{\overline{X}-\mu}{\sigma/\sqrt{n}}}{\sqrt{\frac{S^2}{\sigma^2}}} = \frac{\frac{\overline{X}-\mu}{\sigma/\sqrt{n}}}{\sqrt{\frac{(n-1)/\sigma^2\, S^2}{(n-1)}}} = \frac{Z}{\sqrt{\frac{\chi^2(n-1)}{n-1}}} = \frac{N(0,1)}{\sqrt{\frac{\chi^2(n-1)}{n-1}}}.$$

From (2.3) we see that $T \sim T(n-1)$ has a Student's t-distribution distribution with $n-1$ degrees of freedom. We summarize this result.

Theorem 3.18   (1) If $X \sim N(\mu, \sigma)$ and $X_1,\ldots, X_n$ is a random sample from X, then $T = \dfrac{\overline{X}-\mu}{S/\sqrt{n}}$ has a Student's t-distribution with $n-1$ degrees of freedom.
(2) If X has $EX = \mu$, $Var(X) = \sigma^2$ and $X_1,\ldots, X_n$ is a random sample from X, which may not be normal, then $T = \dfrac{\overline{X}-\mu}{S/\sqrt{n}}$ has an **approximate** Student's t-distribution with $n-1$ degrees of freedom.

Example 3.19   The length of stays at a medical center after heart surgery have mean $\mu = 6.2$ days. A random sample of $n = 20$ such patients resulted in a sample mean of $\bar{x} = 7.8$ with a sample SD of $s = 1.2$. The probability that we have a sample mean of 7.8 or greater is $P(\bar{X} \geq 7.8) = P(\frac{\bar{X}-6.2}{1.2/\sqrt{20}} \geq \frac{7.8-6.2}{1.2/\sqrt{20}}) = P(T(19) \geq 5.963) = \text{tcdf}(5.963, \infty, 19) = 4.86 \times 10^{-6} \approx 0$. This is very unlikely to happen. Maybe μ is incorrect.

Bernoulli Population; p Unknown

If the population is Bernoulli but we don't know the population probability of success p, then we estimate p from the sample proportion given by $\bar{X} = \frac{X_1+\cdots+X_n}{n} = \overline{P}$. We know that

$$E(\overline{P}) = p \quad \text{and} \quad SD(\overline{P}) = \frac{\sqrt{p(1-p)}}{\sqrt{n}}.$$

Everything would be fine except for the fact that we don't know p. We have no choice but to use the information available to us and replace p by the observed sample proportion $\overline{P} = \overline{p}$. The SE is then approximated by $SD(\overline{P}) = \frac{\sqrt{\overline{p}(1-\overline{p})}}{\sqrt{n}}$ and

$$Z = \frac{\overline{P} - p}{\sqrt{\frac{\overline{p}(1-\overline{p})}{n}}} \approx N(0,1) \quad \text{or} \quad \overline{P} \approx N\left(\overline{p}, \sqrt{\frac{\overline{p}(1-\overline{p})}{n}}\right).$$

This should only be used if $n\overline{p} \geq 5$ and $n(1-\overline{p}) \geq 5$.

3.2.1   SAMPLING DISTRIBUTION OF DIFFERENCES OF TWO SAMPLES

In many problems we want to compare the results of two independent random samples from two populations. For those problems we will need the sampling distributions of the differences in the sample means. The following theorem summarizes the results. Since the sample sizes may be different, we use $\overline{X}_n$ to denote the sample mean when the sample size is n.

Theorem 3.20   Let X and Y be two random variables with and let $X_1, X_2, \ldots, X_n$ be a random sample from X and $Y_1, Y_2, \ldots, Y_m$ be an independent random sample from Y. Let $\mu_X = EX, \mu_Y = EY, \sigma_X = SD(X)$, and $\sigma_Y = SD(Y)$.

(a) $E(\overline{X}_n) = \mu_X, SD(\overline{X}_n) = \frac{\sigma_X}{\sqrt{n}}$ and $E(\overline{Y}_m) = \mu_Y, SD(\overline{Y}_m) = \frac{\sigma_Y}{\sqrt{m}}$.

(b) $E(\overline{X}_n - \overline{Y}_m) = \mu_X - \mu_Y$ and $SD(\overline{X}_n - \overline{Y}_m) = \sqrt{\frac{\sigma_X^2}{n} + \frac{\sigma_Y^2}{m}}$.

(c) If $X \sim N(\mu_X, \sigma_X)$, $Y \sim N(\mu_Y, \sigma_Y)$ then $\overline{X}_n - \overline{Y}_m \sim N(\mu_X - \mu_Y, SD(\overline{X}_n - \overline{Y}_m))$.

(d) For large enough n, m, $\overline{X}_n - \overline{Y}_m \approx N(\mu_X - \mu_Y, SD(\overline{X}_n - \overline{Y}_m))$.

(e) If $X \sim$ Bernoulli(p_X), and $Y \sim$ Bernoulli(p_Y), then, if $np_X \geq 5, n(1 - p_X) \geq 5$ and $mp_Y \geq 5, m(1 - p_Y) \geq 5$,

$$\overline{X}_n - \overline{Y}_m \approx N\left(p_X - p_Y, \sqrt{\frac{p_X(1 - p_X)}{n} + \frac{p_Y(1 - p_Y)}{m}}\right).$$

(f) If $X \sim$ Bernoulli(p_X), and $Y \sim$ Bernoulli(p_Y), then, if $np_X \geq 5, n(1 - p_X) \geq 5$ and $mp_Y \geq 5, m(1 - p_Y) \geq 5$,

$$S_n - S_m \approx N\left(np_X - mp_Y, \sqrt{np_X(1 - p_X) + mp_Y(1 - p_Y)}\right).$$

(g) If the sampling is done without replacement from finite populations, the correction factors are used to adjust the SDs.

(h) If the SDs σ_X and σ_Y are unknown, replace the normal distributions with the t−distributions in a similar way as the one-sample cases.

The only point we need to verify is the formulas for the SD of the difference. This follows from independenc:

$$Var(\overline{X}_n - \overline{Y}_m) = Var(\overline{X}_n) + Var(\overline{Y}_m).$$

3.3   PROBLEMS

3.1.   Consider the population box $\boxed{0}\,\boxed{1}\,\boxed{2}\,\boxed{3}\,\boxed{4}$. We will choose random samples of size 2 from this population, **with replacement**.

    (a) Find all samples of size 2. Find the distribution of $\overline{X}$.

    (b) Find $E(\overline{X})$ and $SD(\overline{X})$ using Theorem 3.4 and directly from the first part.

3.2.   Consider the population box $\boxed{7}\,\boxed{1}\,\boxed{2}\,\boxed{3}\,\boxed{4}$. We will choose random samples of size 2 from this population, **with replacement**.

    (a) Find all samples of size 2. Find the distribution of $\overline{X}$.

    (b) Find $E(\overline{X})$ and $SD(\overline{X})$ using Theorem 3.4 and directly from the first part.

    (c) Repeat the first two parts if the samples are drawn **without replacement**.

3.3.   Suppose an investor has 3 types of investments: 20% is at $40, 35% is at $55, and 45% is at $95. If the investor randomly selects 2 of these investments for sale, what is the distribution of the average sale price, i.e., $P(\overline{X} = k)$? Then find $E(\overline{X})$ and $SD(\overline{X})$.

3.4.   The distribution of $\overline{X}$ is given by

k	4	5	6	7	8
$P(\overline{X} = k)$	1/9	2/9	3/9	2/9	1/9

(a) Find $E(\overline{X})$ and $SD(\overline{X})$.

(b) What is the population mean? What is the population σ if the sample size was 9 and sampling was with replacement?

3.5.   A manufacturer has six different devices; device $i = 1, \ldots, 4$ has i defects and devices 5 and 6 have 0 defects. The inspector chooses at random 2 different devices to inspect.

(a) What is the expected total number of defects of the two sample devices? What is the SE for the total?

(b) What is $P(\overline{X} = 3)$?

3.6.   The IQs of 1000 students have an average of 105.5 with an SD of 11. IQs approximately follow a normal distribution. Suppose 150 random samples of size 25 are taken from this population. Find

(a) the mean and SD of $S_{25} = X_1 + \cdots + X_{25}$;

(b) the mean and SD of $\overline{X}$;

(c) the expected number of sample means that are between 98 and 104 inclusive; and

(d) the expected number of sample means < 97.5.

3.7.   Let $X_1, X_2, \ldots, X_{50}$ be a random sample (so independent and identically distributed) with $\mu = 1/4$ and $\sigma = 1/3$. Use the CLT to estimate $P(X_1 + \cdots + X_{50} < 10)$.

3.8.   The mean life of a cell phone is 5 years with an SD of 1 year. Assume the lifetime follows a normal distribution. Find

(a) $P(4.4 < \overline{X} < 5.2)$ when we take a sample of 9 phones;

(b) the 85th percentile of sample means with samples of size 9; and

(c) the probability a cell phone lasts at least 7 years.

3.9.   In the 1965 case of Swain v Alabama, an African-American man appealed to the U.S. Supreme Court his conviction on a charge of rape on the basis of the fact that there were no African-Americans on his jury. At that time in Alabama, only men over the age of 21 were eligible to serve on a jury. Jurors were selected from a panel of 100 ostensibly randomly chosen men chosen from the county. Census data at that time showed that 16% of the eligible members to be selected for the panel were African-Americans. Of the 100 panel members for the Swain jury, 8 were African-American but were not chosen for the jury because of challenges by the attorneys. Use the Central Limit Theorem to approximate the chance that 8 or fewer African-Americans would be on a panel selected at random from the population.

3.10. A random sample $X_1, \ldots, X_{150}$, is drawn from a population with mean $\mu = 25$ and $\sigma = 10$. The population distribution is unknown. Let A be the sample mean of the first 50 and B be the sample mean of the remaining 100.

   (a) What are the approximate distributions of A, B from the CLT?

   (b) Use the CLT to find $P(19 \leq A \leq 26)$ and $P(19 \leq B \leq 26)$.

3.11. An elevator in a hotel can carry a maximum weight of 4000 pounds. The weights of the customers at the hotel are normally distributed with mean 165 pounds and SD $= 12$ pounds. How many passengers can get on the elevator at one time so that there is at most a 1% chance it is overloaded? If the weights are not normally distributed will your answer be approximately correct? Explain.

3.12. A random sample of 25 circuit boards is chosen and their mean life is calculated, $\overline{X}$. The true distribution of the length of life is Exponential with a mean of 5 years. Use the CLT to approximate the probability $P(|\overline{X} - 5| \leq 0.5)$.

3.13. You will flip a fair coin a certain number of times.

   (a) Use both the exact Binomial(n, p) distribution and the normal approximation to find the probabilities $P(X \leq n/2)$, $P(X = n/2)$ and compare the results. Use $n = 10, 20, 30, 40$ and $p = 0.5$. Here X is the number of heads.

   (b) Find $P(X = 2)$ when $n = 4$ using the Binomial and the Normal approximation. Note that $np < 5, n(1 - p) < 5$.

3.14. In a multiple choice exam there are 25 questions each with 4 possible answers only one of which is correct. If a student takes the exam and randomly guesses the answer for each problem, what is the exact and approximate chance the student will get at least 10 correct.

3.15. Five hundred people will each toss a coin which is suspected to be loaded. They will each toss the coin 120 times.

   (a) If it is a fair coin how many people will we expect to get between 40 and 60% heads?

   (b) If in fact 453 people get between 40 and 60% heads, and you declare the coin to be fair, what is the probability you are making a mistake?

3.16. A candidate in an election received 44% of the popular vote. If a poll of a random sample of size 250 is taken find the approximate probability a majority of the sample would be for the candidate.

3.17. Another possible bet in roulette is to bet on a group of 4 numbers; so if any one of the numbers comes up the gambler wins. The bet pays $8 for every $1 bet. Suppose a gambler will play 25 times betting $1 on a group each time.

    (a) What is the population box for this game?

    (b) Use Theorem 3.12 to find the population mean and SD.

    (c) Find the probability of winning 4 or more games.

    (d) Find the approximate probability of coming out ahead.

3.18. Prove Theorem 3.12.

3.19. Let T_n denote a Student's t rv with n degrees of freedom. Find

    (a) $P(T_7 < 2.35)$.   (c) $P(-1.45 < T_{12} < 2.2)$.

    (b) $P(T_{22} > 1.33)$.   (d) the number c so that $P(T_{12} > c) = 0.95$.

3.20. Let $Y \sim \chi^2(8)$.

    (a) Find EY and $Var(Y)$.   (c) Suppose $P(a < Y < b) = 0.8$. Find a, b so that the two tails have the same area.

    (b) Suppose $P(Y > a) = 0.05, P(Y < b) = 0.1, P(Y < c) = 0.9, P(Y > d) = 0.95$. Find a, b, c, and d.   (d) Suppose $P(\chi^2(1) < a) = 0.23$, and $P(Z^2 < b) = 0.23$. Find a, b and determine if $a = b^2$.

3.21. A random sample of 400 people was taken from the population of factory workers. 213 worked for factories with 100 or more employees. Find the probability of getting a sample proportion of 0.5325 or more, when the true proportion is known to be 0.51.

3.22. Ten measurements of the diameter of a ball resulted in a sample mean $\bar{x} = 4.38$ cm with sample $SD = 0.08$. Given that the mean diameter should be $\mu = 4.30$ find the probability $P(|\overline{X} - 4.30| > 0.08)$.

3.23. The heights of a population of 300 male students at a college are normally distributed with mean 68 inches. Suppose 80 male students are chosen at random and their sample average is 66 inches with a sample SD of 3 inches. What are the chances of drawing such a sample with $\bar{x} \leq 66$ if the sample is drawn with or without replacement?

3.24. A real estate office takes a random sample of 35 rental units and finds a sample average rent paid of $1,200 with a sample SD of $325. The rents do not follow the normal curve. Find the probability the sample average would be $1,200 or more if the actual mean population rent is $1,250.

3.25. A random sample of 1000 people is taken to estimate the percentage of Republicans in the population. 467 people in the sample claim to be Republicans. Find the chance that the percentage of Republicans in a sample of size 1000 will be in the range from 45% to 48%.

3.26. Three organizations take a random sample of size 40 from a Bernoulli population with unknown p. The number of 1s in the first sample is 8, in the second sample is 10, and the third sample is 13. Find the estimated population proportion p for each sample along with the SE for each sample.

3.27. Given a random sample $X_1, \ldots X_n$ from a population with $E(X) = \mu$, $Var(X) = \sigma^2$, find the mean and variance of the estimator $\hat{\mu} = (X_1 + X_n)/2$.

3.28. A random sample of 20 measurements will be taken of the pressure in a chamber. It is assumed that the measurements will be normally distributed with mean 0 and variance $\sigma^2 = 0.004$. What are the chances the sample mean will be within 0.01 of the true mean?

3.29. A measurement process is normally distributed with known variance $\sigma^2 = .054$ but unknown mean. What sample size is required in order to be sure the sample mean is within 0.01 of the true mean with probability at least 0.95?

3.30. The time to failure of a cell phone's battery is 2.78 hours, where failure means a low battery indicator will flash.

  (a) Given that 100 measurements are made with a sample SD of $s = 0.26$, what is the probability the sample mean will be within 0.05 of the true mean?

  (b) How many measurements need to be taken to ensure $P(|\overline{X} - \mu| < 0.05) \geq 0.98$? Assume $s = 0.26$ is a good approximation to σ.

3.31. Suppose we have 49 data values with sample mean $\overline{x} = 6.25$ and sample SD 6. Find the probability of obtaining a sample mean of 6.25 or greater if the true population mean is $\mu = 4$.

3.32. Two manufacturers each produce a semiconductor component with a mean lifetime of $\mu_X = 1400$ hours and $\sigma_X = 200$ for company X, and $\mu_Y = 1200$, $\sigma_Y = 100$ for company Y. Suppose 125 components from each company are randomly selected and tested.

  (a) Find the probability that X's components will have a mean lifetime at least 160 hours longer than Y's.

(b) Find the probability that X's components will have a mean lifetime at least 250 hours longer than Y's.

3.33. Two drug companies, A and B, have competing drugs for a disease. Each companies drug has a 50% chance of a cure. They will each choose 50 patients at random and administer their drug. Find the approximate probability company A will achieve 5 or more cures than company B.

3.34. The population mean score of students on an exam is 72 with an SD of 6. Suppose two groups of students are independently chosen at random with 26 in one group and 32 in the other.

(a) Find the probability the sample means will differ by more than 3.

(b) Find the probability the sample means will differ by at least 2 but no more than 4 points.

3.35. Post election results showed the winning candidate with 53% of the vote. Find the probability the two independent random samples with 200 voters each would indicate a difference of more than 12% in their two voting proportions.

3.36. Let $\{Z_1, \ldots, Z_{16}\}$ be a random sample from $N(0, 1)$ and $\{X_1, \ldots, X_{64}\}$ an independent random sample from $X \sim N(\mu, 1)$.

(a) Find $P(Z_{12} > 2)$.

(b) Find $P\left(\sum_{i=1}^{16} Z_i > 2\right)$.

(c) Find $P\left(\sum_{i=1}^{16} Z_i^2 > 16\right)$.

(d) Find a value of c such that

$$P\left(\frac{1}{15}\sum_{i=1}^{16}(Z_i - \overline{Z})^2 > c\right) = 0.05.$$

(e) Find the distribution of

$$Y = \sum_{i=1}^{16} Z_i^2 + \sum_{i=1}^{64}(X_i - \mu)^2.$$

CHAPTER 4

Confidence and Prediction Intervals

This chapter introduces the concept of a **statistical interval**. The principal types of statistical intervals are **confidence intervals** and **prediction intervals**. Instead of using a single number (a point estimate) to estimate quantities like the mean, a statistical interval includes the point estimate and an interval around it (an interval estimate) quantifying the errors involved in the estimate.

4.1 CONFIDENCE INTERVALS FOR A SINGLE SAMPLE

We will use ϑ to denote a parameter of interest in the pmf or pdf which we are trying to estimate. Normally this is the population mean μ or SD σ, or the population binomial proportion p. The following is the precise definition of what it means to be a confidence interval for the parameter ϑ. We give the definition for a continuous random variable X with pdf $f_X(x; \vartheta)$. The definition if X is discrete is similar.

Definition 4.1 Let $X_1, X_2, \ldots, X_n$ be a random sample from a random variable X with pdf $f_X(x; \vartheta)$. Given $0 < \alpha < 1$, a $100(1 - \alpha)\%$ **confidence interval** for ϑ is an open interval with **random endpoints** of the form $l(X_1, X_2, \ldots, X_n)$ and $u(X_1, X_2, \ldots, X_n)$ such that

$$\boxed{P\left(l\left(X_1, X_2, \ldots, X_n\right) < \vartheta < u\left(X_1, X_2, \ldots, X_n\right)\right) = 1 - \alpha.}$$

The percentage $100(1 - \alpha)\%$ is called the **confidence level** of the interval.

In the remainder of the book we will use **CI** to denote Confidence Interval.

Remark 4.2 The value of α sets the confidence level, and α represents the probability the interval (l, u) does **not** contain the parameter ϑ. The probability that the value of the unknown parameter ϑ lies in the interval can be adjusted depending on our choice of α. For example, if we would like the probability to be 0.99 (a confidence level of 99%), then we would choose $\alpha = 0.01$. If it is acceptable that the probability be only 0.90 (a confidence level of 90%), then we can instead choose $\alpha = 0.10$. We would expect that the smaller the value of α, meaning a higher confidence level, the wider the interval and vice versa. That is, the more confidence we want, the less we have to be confident of. No matter what the level of confidence (except for $\alpha = 0$), the random interval (l, u) may not contain the true value of ϑ.

Remark 4.3   The interval in the definition has random endpoints which depend on the random sample. How do we obtain an actual interval of real numbers? If $X_1 = x_1$, $X_2 = x_2$, and $X_n = x_n$ is the set of actual observed sample values,[1] then an interval on the real line is obtained by evaluating the functions l and u at the sample values. That is, the observed sample endpoints of the interval are $l = l(x_1, x_2, \ldots, x_n)$ and $u = u(x_1, x_2, \ldots, x_n)$. Then the **observed** $100(1 - \alpha)\%$ confidence interval is simply (l, u). (Although $(l(x_1, \ldots, x_n), u(x_1, \ldots, x_n))$ does not have random endpoints, we will still refer to it as a confidence interval. It is important to keep in mind however that $P(l(X_1, \ldots, X_n) < \vartheta < u(X_1, \ldots, X_n))$ has meaning, but $P(l(x_1, \ldots, x_n) < \vartheta < u(x_1, \ldots, x_n))$ does not since there are no random variables present in the latter expression.)

The true value of ϑ is either in $(l(x_1, \ldots, x_n), u(x_1, \ldots, x_n)) = (l, u)$ or it is not. If (l, u) does contain ϑ, let us consider that a **success**. A **failure** occurs when ϑ is not in the interval. We may consider this a Bernoulli rv with $1 - \alpha$ as the probability of success. If we construct N $100(1 - \alpha)\%$ confidence intervals by taking N observations of the random sample, the rv S which counts the number of successes is $S \sim \text{Binom}(N, 1 - \alpha)$. Then, we expect $E(S) = N(1 - \alpha)$ of these constructed intervals to actually be successes, i.e., contain ϑ. For instance, if $\alpha = 0.05$, and we perform the experiment 100 times, we expect 95 of the intervals to contain the true value of ϑ and 5 not to contain ϑ. Of course, when we obtain data and construct the confidence interval, we will not know whether this particular interval contains ϑ or not, but we have $100(1 - \alpha)\%$ **confidence** that it does.

4.1.1   CONTROLLING THE ERROR OF AN ESTIMATE USING CONFIDENCE INTERVALS

An important feature of confidence intervals is that they allow us to control the error of the estimates produced by estimators. Specifically, suppose that $\vartheta_e = \hat{\vartheta}(x_1, \ldots, x_n)$ is an estimate obtained by substituting observed sample values into an estimator $\hat{\vartheta}(X_1, \ldots, X_n)$ for the parameter ϑ.

Definition 4.4   The **error** ε of the estimate ϑ_e is the quantity $\varepsilon = |\vartheta_e - \vartheta|$.

The error of the estimate is therefore the amount by which the estimate deviates from the true value of ϑ.

---

[1]Capital values, X, are random variables, while small script, x, is the observed value of the rv, X.

4.1.2   PIVOTAL QUANTITIES

How do we construct confidence intervals for parameters in probability distributions? Normally we need what are called **pivotal quantities**, or **pivots** for short.

Definition 4.5   Let $X_1, X_2, \ldots, X_n$ be a random sample from a random variable X with pdf $f_X(x; \vartheta)$. A **pivotal quantity** for ϑ is a random variable $h(X_1, X_2, \ldots, X_n, \vartheta)$ whose distribution does not depend on ϑ, i.e., it is the same for every value of ϑ.

The function h is a random variable constructed from the random sample **and** the constant ϑ. Here's an example of a pivotal quantity.

Example 4.6   Take a sample X of size one from a normal distribution with unknown μ but known σ. The rv

$$Z = \frac{X - \mu}{\sigma} \sim N(0, 1)$$

is a standard normal random variable that doesn't depend on the value of μ. Therefore, $Z = (X - \mu)/\sigma$ is a **pivotal quantity for** μ. These quantities will also be known as **test statistics** in a later context.

Pivotal Quantities for the Normal Parameters

Let $X_1, X_2, \ldots, X_n$ be a random sample from a normal random variable $X \sim N(\mu, \sigma)$. The following distributions leading to pivotal quantities were introduced in Chapter 3.

$\sigma^2 = \sigma_0^2$ **is known**
Since $E(\overline{X}) = \mu$ and $SD(\overline{X}) = \sigma_0/\sqrt{n}$, a pivotal quantity for μ is

$$\boxed{\frac{\overline{X} - \mu}{\sigma_0/\sqrt{n}} \sim N(0, 1).}$$

σ^2 **is unknown**
In this case, σ has to be estimated from the sample. Therefore, a pivotal quantity for μ is

$$\boxed{\frac{\overline{X} - \mu}{S_X/\sqrt{n}} \sim t(n - 1).}$$

We now turn to finding a pivotal quantity for the variance. We again consider two cases.

Table 4.1: Pivotal quantities for $N(\mu, \sigma)$

Parameter	Conditions	Pivotal Quantity	Distribution
μ	$\sigma = \sigma_0$ known	$\dfrac{\overline{X} - \mu}{\sigma_0/\sqrt{n}}$	$N(0, 1)$
μ	σ unknown	$\dfrac{\overline{X} - \mu}{S/\sqrt{n}}$	$t(n - 1)$
σ^2	$\mu = \mu_0$ known	$\dfrac{1}{\sigma^2} \sum\limits_{i=1}^{n} (X_i - \mu_0)^2$	$\chi^2(n)$
σ^2	μ unknown	$\dfrac{1}{\sigma^2} \sum\limits_{i=1}^{n} (X_i - \overline{X})^2$	$\chi^2(n - 1)$

$\mu = \mu_0$ **is known**

In this case,

$$\sum_{i=1}^{n} \left(\frac{X_i - \mu_0}{\sigma} \right)^2 \sim \chi^2(n).$$

Recall that this follows from the fact that each term $(X_i - \mu)/\sigma \sim N(0, 1)$, and the sum of n squares of independent standard normals is $\chi^2(n)$.

μ **is unknown**

In this case, μ has to be estimated from the sample, and the pivot becomes

$$\frac{1}{\sigma^2} \sum_{i=1}^{n} (X_i - \overline{X})^2 \sim \chi^2(n - 1).$$

There is a loss of a degree of freedom due to estimating μ. We summarize our results in Table 4.1.

Remark 4.7   A TI program to obtain critical values of the χ^2 distribution is given in Remark 2.67.

4.1.3   CONFIDENCE INTERVALS FOR THE MEAN AND VARIANCE OF A NORMAL DISTRIBUTION

Now that we have pivotal quantities for the parameters of interest, we can proceed to construct confidence intervals for the normal random variable.

A Confidence Interval for the Mean, σ known

We first construct a $100(1-\alpha)\%$ confidence interval for the mean μ of a normal random variable $X \sim N(\mu, \sigma)$ when $\sigma = \sigma_0$ is known. Let $X_1, X_2, \ldots, X_n$ be a random sample. We know that $Z = \frac{\overline{X}-\mu}{\sigma_0/\sqrt{n}} \sim N(0, 1)$. Therefore, we want

$$P\left(-z_{\alpha/2} < Z < z_{\alpha/2}\right) = P\left(-z_{\alpha/2} < \frac{\overline{X}-\mu}{\sigma_0/\sqrt{n}} < z_{\alpha/2}\right) = 1 - \alpha,$$

where $z_{\alpha/2} = \text{invNorm}(1 - \alpha/2)$ is the $\alpha/2$ critical value of Z. Rearranging we get

$$P\left(\overline{X} - z_{\alpha/2}\frac{\sigma_0}{\sqrt{n}} < \mu < \overline{X} + z_{\alpha/2}\frac{\sigma_0}{\sqrt{n}}\right) = 1 - \alpha.$$

The confidence interval with random endpoints is therefore

$$(l(X_1, \ldots, X_n), u(X_1, \ldots, X_n)) = \left(\overline{X} - z_{\alpha/2}\frac{\sigma_0}{\sqrt{n}}, \overline{X} + z_{\alpha/2}\frac{\sigma_0}{\sqrt{n}}\right).$$

For observed values of $X_1 = x_1, \ldots, X_n = x_n$, the confidence interval is

$$\left(\overline{x} - z_{\alpha/2}\frac{\sigma_0}{\sqrt{n}}, \overline{x} + z_{\alpha/2}\frac{\sigma_0}{\sqrt{n}}\right).$$

We are $100(1-\alpha)\%$ confident that the true value of μ is in this interval. If μ really is in this interval, then the error is

$$\varepsilon = |\overline{x} - \mu| \le z_{\alpha/2}\frac{\sigma_0}{\sqrt{n}}.$$

But μ might not be in the interval, and so we can only say that we are $100(1-\alpha)\%$ confident that the error is no more than $z_{\alpha/2}\sigma_0/\sqrt{n}$.

Remark 4.8 Looking at the error ε, we see that ε can be decreased in several ways. For a fixed sample size n, the larger α is, i.e., the lower the confidence level, the smaller ε will be. This is because a smaller confidence level implies a smaller value of $z_{\alpha/2}$. On the other hand, for a fixed α, if we increase the sample size, ε gets smaller. For a given confidence level, the only way to decrease the error is to increase the sample size. Notice that as $n \to \infty$, i.e., as the sample size becomes larger and larger, the error shrinks to zero.

Example 4.9 A sample $X_1, X_2, \ldots, X_{106}$ of 106 healthy adults have their temperatures taken during a routine physical checkup, and it is found that the mean body temperature of the sample is $\overline{x} = 98.2°\text{F}$. Previous studies suggest that $\sigma_0 = 0.62°\text{F}$. Assume the body temperatures are drawn from a normal population. Set $\alpha = 0.05$ for a confidence level of 95%. A 95% confidence interval for μ is, since $z_{.025} = \text{invNorm}(.975) = 1.96$,

$$\left(\bar{x} - z_{.025} \frac{\sigma_0}{\sqrt{106}}, \bar{x} + z_{.025} \frac{\sigma_0}{\sqrt{106}} \right) = (98.08°, 98.32°).$$

We are 95% confident that the true mean healthy adult body temperature is between 98.08°F and 98.32°F. The traditional value of 98.6°F for the body temperature of a healthy adult is not in this interval! There is a 5% chance that the interval missed the true value, but this result provides some evidence that the true value may not be 98.6°F. Finally, we note that

$$\varepsilon = |\bar{x} - \mu| \leq z_{.025} \frac{\sigma_0}{\sqrt{106.0}} = 0.12°F.$$

Therefore, we are 95% confident that our estimate of $\bar{x} = 98.2°F$ as the mean healthy adult body temperature deviates from the true temperature by at most 0.12°F.

Sample Size for Given Level of Confidence

For a fixed level of confidence, the only parameter we can control is the sample size n. What we want to know is how large the sample size needs to be so that we can guarantee the error is no greater than some given amount, say d. We would need to require that

$$\frac{z_{\alpha/2}\sigma_0}{\sqrt{n}} \leq d.$$

Solving the inequality for n, we obtain

$$z_{\alpha/2}\sigma_0 \leq d\sqrt{n} \implies \sqrt{n} \geq \frac{z_{\alpha/2}\sigma_0}{d} \implies \boxed{n = \left\lceil \left(\frac{z_{\alpha/2}\sigma_0}{d} \right)^2 \right\rceil}.$$

The smallest sample size that will do the job is given by the smallest integer greater than or equal to $(z_{\alpha/2}\sigma_0/d)^2$.

Example 4.10   Going back to the previous example, suppose that we take $d = 0.05$, and we want to be 99% percent confident that our estimate $\bar{x}$ differs from the true value of the mean healthy adult body temperature by at most 0.05°F. In this case,

$$n = \left\lceil \left(\frac{z_{\alpha/2}\sigma_0}{d} \right)^2 \right\rceil = \left\lceil \left(\frac{z_{0.005}0.62}{0.05} \right)^2 \right\rceil = \left\lceil \left(\frac{(2.58)(0.62)}{0.05} \right)^2 \right\rceil = 1{,}024.$$

We would have to take a much larger sample than the original $n = 106$ to be 99% confident that our estimate is within 0.05°F of the true value.

Now we have to work on the more realistic problem that the variance of the population is unknown.

A Confidence Interval for the Mean, σ Unknown

If the variance σ^2 is unknown, we have no choice but to replace σ with its estimate S_X from the sample. A pivotal quantity for μ is given by $T = \dfrac{\overline{X} - \mu}{S_X/\sqrt{n}} \sim t(n - 1)$. Repeating essentially the same derivation as before, we obtain the $100(1 - \alpha)\%$ confidence interval for μ as

$$\left(\overline{x} - t(n - 1, \alpha/2)\frac{s_X}{\sqrt{n}}, \overline{x} + t(n - 1, \alpha/2)\frac{s_X}{\sqrt{n}} \right).$$

Just as $\overline{x}$ is the sample mean estimate for μ, s_X is the sample SD estimate for σ.

Example 4.11   During a certain two-week period during the summer, the number of drivers speeding on Lake Shore Drive in Chicago is recorded. The data values are listed below.

Drivers Speeding on LSD						
10	15	11	9	12	7	10
6	15	12	8	12	15	9

Assume the population is normal, meaning that the number of drivers speeding in a two-week period is normally distributed. The variance is unknown and must be estimated. We compute $\overline{x} = 10.93$ and $s_X^2 = 10.07$ from the data values. Set $\alpha = 0.05$ for a confidence level of 95%. A 95% confidence interval for μ, the true mean number of drivers speeding on a given day, is given by

$$\left(\overline{x} - t(13, 0.025)\frac{s_X}{\sqrt{14}}, \overline{x} + t(13, 0.025)\frac{s_X}{\sqrt{14}} \right) = (9.10, 12.47).$$

A Confidence Interval for the Variance

We now turn to the variance. We consider two cases: the mean $\mu = \mu_0$ is known, and μ is unknown.

First consider the case when $\mu = \mu_0$ **is known**. From our table of pivotal quantities, we have that

$$\frac{1}{\sigma^2} \sum_{i=1}^{n} (X_i - \mu_0)^2 \sim \chi^2(n).$$

If we set our confidence level at $100(1 - \alpha)\%$, then

$$P\left(\chi^2(n, 1 - \alpha/2) < \frac{1}{\sigma^2} \sum_{i=1}^{n} (X_i - \mu_0)^2 < \chi^2(n, \alpha/2) \right) = 1 - \alpha.$$

Notice that the χ^2 distribution is not symmetric and so we need to consider two distinct χ^2 critical values. For instance, $100\alpha/2\%$ of the area under the χ^2 pdf is to the right of $\chi^2(n, \alpha/2)$, and $100\alpha/2\%$ of the area under the χ^2 pdf is to the left of $\chi^2(n, 1 - \alpha/2)$. Solving this inequality for σ^2, we get

$$P\left(\frac{\sum\limits_{i=1}^{n}(X_i - \mu_0)^2}{\chi^2(n, \alpha/2)} < \sigma^2 < \frac{\sum\limits_{i=1}^{n}(X_i - \mu_0)^2}{\chi^2(n, 1 - \alpha/2)}\right) = 1 - \alpha.$$

A $100(1 - \alpha)\%$ confidence interval for σ^2 is therefore given by

$$\boxed{\left(\frac{\sum\limits_{i=1}^{n}(x_i - \mu_0)^2}{\chi^2(n, \alpha/2)}, \frac{\sum\limits_{i=1}^{n}(x_i - \mu_0)^2}{\chi^2(n, 1 - \alpha/2)}\right).}$$

In a similar way, if the **mean μ is unknown**, then it must be estimated as $\overline{X}$. This time our table of pivotal quantities gives

$$\frac{1}{\sigma^2}\sum_{i=1}^{n}\left(X_i - \overline{X}\right)^2 \sim \chi^2(n - 1).$$

We write $\sum\limits_{i=1}^{n}\left(X_i - \overline{X}\right)^2 = (n - 1)S_X^2$. In this case, a $100(1 - \alpha)\%$ confidence interval for σ^2 is given by

$$\boxed{\left(\frac{(n - 1)s_X^2}{\chi^2(n - 1, \alpha/2)}, \frac{(n - 1)s_X^2}{\chi^2(n - 1, 1 - \alpha/2)}\right).}$$

We summarize all our confidence intervals for the parameters of the normal random variable in Table 4.2.

4.1.4   CONFIDENCE INTERVALS FOR A PROPORTION

We now turn to Bernoulli(p) populations which are either successes or failures, i.e., each individual in the population is either a 0 or a 1, and we are interested in estimating the percentage of 1s in the population, which is p.

Let $X \sim \text{Binom}(n, p)$. Recall that X may be represented as

$$X = X_1 + X_2 + \cdots + X_n$$

where each X_i is a Bernoulli(p) random variable. Recall also that $E(X) = np$ and $Var(X) = np(1 - p)$. If n is large enough, the Central Limit Theorem can be invoked to approximate the random variable

$$\frac{X - np}{\sqrt{np(1 - p)}} \sim N(0, 1).$$

The normal approximation is appropriate if $np > 5$ and $n(1 - p) > 5$.

Table 4.2: Confidence intervals for the parameters of the normal distribution

Parameter	Conditions	Confidence Interval
μ	$\sigma = \sigma_0$ known	$\left(\overline{x} - z_{\alpha/2}\dfrac{\sigma}{\sqrt{n}}, \overline{x} + z_{\alpha/2}\dfrac{\sigma}{\sqrt{n}}\right)$
μ	σ unknown	$\left(\overline{x} - t(n-1, \alpha/2)\dfrac{s_X}{\sqrt{n}}, \overline{x} + t(n-1, \alpha/2)\dfrac{s_X}{\sqrt{n}}\right)$
σ^2	$\mu = \mu_0$ known	$\left(\dfrac{\sum\limits_{i=1}^{n}(x_i - \mu_0)^2}{\chi^2(n, \alpha/2)}, \dfrac{\sum\limits_{i=1}^{n}(x_i - \mu_0)^2}{\chi^2(n, 1 - \alpha/2)}\right)$
σ^2	μ unknown	$\left(\dfrac{(n-1)s_X^2}{\chi^2(n-1, \alpha/2)}, \dfrac{(n-1)s_X^2}{\chi^2(n-1, 1 - \alpha/2)}\right)$

Confidence Interval for p

Using the normal approximation, we have

$$P\left(-z_{\alpha/2} < \frac{X - np}{\sqrt{np(1-p)}} < z_{\alpha/2}\right) \cong 1 - \alpha,$$

and so

$$P\left(-z_{\alpha/2}\sqrt{np(1-p)} < X - np < z_{\alpha/2}\sqrt{np(1-p)}\right) \cong 1 - \alpha.$$

Dividing through by n, we get

$$P\left(-z_{\alpha/2}\sqrt{\frac{p(1-p)}{n}} < \overline{X} - p < z_{\alpha/2}\sqrt{\frac{p(1-p)}{n}}\right) \cong 1 - \alpha.$$

Solve this inequality for p to get

$$P\left(\overline{X} - z_{\alpha/2}\sqrt{\frac{p(1-p)}{n}} < p < \overline{X} + z_{\alpha/2}\sqrt{\frac{p(1-p)}{n}}\right) \cong 1 - \alpha.$$

The problem with this interval is that the endpoints contain the unknown parameter p. To eliminate p from the endpoints, we use the fact that p can be approximated by $\overline{X}$. If we take this approach, called the **bootstrap method**, we obtain

$$P\left(\overline{X} - z_{\alpha/2}\sqrt{\frac{\overline{X}(1-\overline{X})}{n}} < p < \overline{X} + z_{\alpha/2}\sqrt{\frac{\overline{X}(1-\overline{X})}{n}}\right) \cong 1 - \alpha.$$

The $100(1 - \alpha)\%$ confidence interval for p using this approach is

$$\left(\overline{p} - z_{\alpha/2} \sqrt{\frac{\overline{p}(1 - \overline{p})}{n}}, \ \overline{p} + z_{\alpha/2} \sqrt{\frac{\overline{p}(1 - \overline{p})}{n}} \right).$$

We have replaced the random variable $\overline{X}$ with the observed sample proportion $\overline{p} = \overline{x}$. The center of the confidence interval is $\overline{p}$ which is our approximation for p.

Error Bounds and Sample Size.

For the population proportion, the error is $\varepsilon = |\overline{p} - p|$, the amount by which the estimate $\overline{p}$ deviates from the true value of p. An error bound is easily obtained since we are $100(1 - \alpha)\%$ confident that

$$\varepsilon \leq z_{\alpha/2} \sqrt{\frac{\overline{p}(1 - \overline{p})}{n}}.$$

Now suppose we wish to construct an interval for which we are $100(1 - \alpha)\%$ confident that $\varepsilon \leq d$ for some specified $d > 0$. Using the fact that $\overline{p}(1 - \overline{p}) \leq 1/4$, since the function $f(x) = x(1 - x)$ achieves a maximum value of $1/4$ on the interval $[0, 1]$, we obtain

$$z_{\alpha/2} \sqrt{\frac{\overline{p}(1 - \overline{p})}{n}} \leq z_{\alpha/2} \sqrt{\frac{1}{4n}}.$$

Therefore

$$z_{\alpha/2} \sqrt{\frac{1}{4n}} \leq d \implies \boxed{n \geq \left\lceil \frac{z_{\alpha/2}^2}{4d^2} \right\rceil}.$$

The value of n is a sample size for which we can be $100(1 - \alpha)\%$ confident that the error $\varepsilon \leq d$. For example, if $\alpha = 0.05$ and $d = 0.01$, the sample size for which $\varepsilon \leq 0.01$ is $n \geq \lceil 1.96^2/(4(0.01)^2) \rceil = 9{,}604$.

Remark 4.12   The estimate of the sample size $n \geq \left\lceil z_{\alpha/2}^2/(4d^2) \right\rceil$ is the conservative estimate because we have replaced the unknown $\overline{p}$ with $1/2$. Another method of estimating n is to run a two-stage experiment. In the first stage, an arbitrary sample size $n \geq 30$ is taken, and the estimate $\overline{p}$ is computed. Then the sample size to obtain an error bound of d is calculated by

$$\boxed{n \geq \left\lceil \overline{p}(1 - \overline{p}) \left(\frac{z_{\alpha/2}}{d} \right)^2 \right\rceil}.$$

Example 4.13   A gardener is trying to grow a rare species of orchid in a greenhouse. Let X denote the number of plants that survive under greenhouse conditions. Of the (random sample of) 50 plants she originally potted, only 17 survived. The random variable $X \sim \text{Binom}(50, p)$ has an unknown p. Suppose we wish to compute a 95% confidence interval for the percentage p of plants that will survive in the greenhouse. Notice that our estimate for p is $\overline{p} = 0.34$.

For the endpoints of the confidence interval, we obtain

$$l = 0.34 - z_{0.025} \sqrt{\frac{0.34(1 - 0.34)}{50}} = 0.209 \text{ and } u = 0.34 + z_{0.025} \sqrt{\frac{0.34(1 - 0.34)}{50}} = 0.471 .$$

The 95% confidence interval is $(0.209, 0.471)$. This time $\varepsilon \leq 0.471 - 0.34 = 0.131$. If we want $\varepsilon \leq 0.09 = d$, then we need a sample of size

$$n \geq \left\lceil .34(1 - .34) \left(\frac{z_{.025}}{.09} \right)^2 \right\rceil = \lceil 106.422 \rceil = 107.$$

If we had not taken a sample of $n = 50$ and didn't have an estimate of $\overline{p} = .34$, we would need a sample of size $n \geq \lceil \frac{1}{4} \left(\frac{z_{.025}}{.09} \right)^2 \rceil = 119$ orchids to guarantee $\varepsilon \leq 0.09$.

4.1.5   ONE-SIDED CONFIDENCE INTERVALS

Sometimes we only desire an upper or lower bound on the unknown parameter ϑ of a probability distribution at a given confidence level. In this case we can generate **one-sided** confidence intervals. For example, only an upper confidence bound on a speed limit is needed to ensure the maximum speed is not exceeded. A lower confidence bound is needed to ensure some minimum design specification is met. The intervals described so far have all been **two-sided**.

Definition 4.14   Let $X_1, X_2, \ldots, X_n$ be a random sample from a random variable X with pdf $f_X(x; \vartheta)$. A $100(1 - \alpha)\%$ **one-sided confidence interval** for ϑ is an interval of the form $(l(X_1, X_2, \ldots, X_n), +\infty)$ or $(-\infty, u(X_1, X_2, \ldots, X_n))$ such that

$$P(l(X_1, X_2, \ldots, X_n) < \vartheta) = 1 - \alpha \quad \text{or} \quad P(\vartheta < u(X_1, X_2, \ldots, X_n)) = 1 - \alpha,$$

respectively. The quantity $100(1 - \alpha)\%$ is called the **confidence level** of the respective intervals.

One-sided confidence intervals can be constructed for all the parameters in this chapter. It will suffice to give a simple example to illustrate the process of constructing such an interval.

Example 4.15   Consider Example 4.11 involving speeders on Lake Shore Drive discussed previously. Suppose we wish to construct 95% one-sided confidence intervals for the mean number of speeders. As before, if the variance σ^2 is unknown, then a pivotal quantity for μ is $\frac{\overline{X} - \mu}{S_X / \sqrt{n}} \sim t(n - 1)$. To obtain an **upper one-sided confidence interval**, we set

$$P\left(\frac{\overline{X} - \mu}{S_X / \sqrt{n}} > -t(n - 1, \alpha) \right) = 1 - \alpha,$$

or equivalently,

$$P\left(\mu < \overline{X} + t(n-1,\alpha)S_X/\sqrt{n}\right) = 1 - \alpha.$$

A $100(1-\alpha)\%$ upper confidence interval for μ is $(-\infty, \overline{X} + t(n-1,\alpha)S_X/\sqrt{n})$. An upper $100(1-\alpha)\%$ one-sided observed confidence interval is given by

$$\left(-\infty, \overline{x} + t(n-1,\alpha)s_X/\sqrt{n}\right).$$

For our example, if $\alpha = 0.05$, $(-\infty, 10.93 + (1.77)(3.17)/\sqrt{14}) = (-\infty, 12.43)$ is a 95% upper confidence interval for μ. We would say that we are 95% confident that the mean number of speeders is no more than 12.43. If someone claimed the mean number of speeders was at least 15, our upper bound would be good evidence against that.

Similarly, a lower $100(1-\alpha)\%$ one-sided observed confidence interval is given by

$$\left(\overline{x} - t(n-1,\alpha)s_X/\sqrt{n}, +\infty\right).$$

For our example, we obtain $(10.93 - (1.77)(3.17)/\sqrt{14}, +\infty) = (9.43, +\infty)$ as a 95% lower confidence interval for μ. We say that we are 95% confident that the mean number of speeders is at least 9.43. By contrast, the two-sided 95% confidence interval for the mean number of speeders is $(9.10, 12.47)$, and we are 95% confident the mean number of speeders is in that interval.

One-sided intervals are not constructed simply by taking the upper and lower values of a two-sided interval because in the two-sided case we use $z_{\alpha/2}$, but we use z_α in the one-sided case.

4.2   CONFIDENCE INTERVALS FOR TWO SAMPLES

In order to compare two populations we need to find a confidence interval for the differences in the means, ratio of the variances, or difference of proportions.

4.2.1   DIFFERENCE OF TWO NORMAL MEANS

Let $X_1, X_2, \ldots, X_m$ and $Y_1, Y_2, \ldots, Y_n$ be **independent** random samples from two **normal populations** $X \sim N(\mu_X, \sigma_X)$ and $Y \sim N(\mu_Y, \sigma_Y)$, respectively. We will derive $100(1-\alpha)\%$ confidence intervals for the difference $\mu_X - \mu_Y$ under three different sets of conditions on the variances of these populations:

- σ_X^2 and σ_Y^2 both known,

- σ_X^2 and σ_Y^2 both unknown but equal, and

- σ_X^2 and σ_Y^2 both unknown and unequal.

Observe that we do not require the same sample size for the two populations. We will denote the sample averages by $\overline{X}_m$ and $\overline{Y}_n$ to emphasize the sample sizes.

Variances of Both Populations Known

We first assume that $\sigma_X^2 = \sigma_{0,X}^2$ and $\sigma_Y^2 = \sigma_{0,Y}^2$ are both known. By independence, $Var(\overline{X}_m - \overline{Y}_n) = \frac{\sigma_{0,X}^2}{m} + \frac{\sigma_{0,Y}^2}{n}$, and

$$\overline{X}_m - \overline{Y}_n \sim N\left(\mu_X - \mu_Y, \sqrt{\frac{\sigma_{0,X}^2}{m} + \frac{\sigma_{0,Y}^2}{n}}\right).$$

Consequently, the random variable $\frac{(\overline{X}_m - \overline{Y}_n) - (\mu_X - \mu_Y)}{\sqrt{\frac{\sigma_{0,X}^2}{m} + \frac{\sigma_{0,Y}^2}{n}}} = Z \sim N(0, 1)$ is a pivotal quantity.

Therefore, we have

$$P\left(-z_{\alpha/2} < \frac{(\overline{X}_m - \overline{Y}_n) - (\mu_X - \mu_Y)}{\sqrt{\frac{\sigma_{0,X}^2}{m} + \frac{\sigma_{0,Y}^2}{n}}} < z_{\alpha/2}\right) = 1 - \alpha.$$

For simplicity, set $D_{n,m} = \sqrt{\frac{\sigma_{0,X}^2}{m} + \frac{\sigma_{0,Y}^2}{n}}$. Then after some algebra we get

$$P\left((\overline{X}_m - \overline{Y}_n) - z_{\alpha/2}D_{n,m} < \mu_X - \mu_Y < (\overline{X}_m - \overline{Y}_n) + z_{\alpha/2}D_{n,m}\right) = 1 - \alpha.$$

The $100(1 - \alpha)\%$ confidence interval for $\mu_X - \mu_Y$ in the case when both variances are known is therefore given by

$$\boxed{\left((\bar{x}_m - \bar{y}_n) - z_{\alpha/2}\sqrt{\frac{\sigma_{0,X}^2}{m} + \frac{\sigma_{0,Y}^2}{n}}, (\bar{x}_m - \bar{y}_n) + z_{\alpha/2}\sqrt{\frac{\sigma_{0,X}^2}{m} + \frac{\sigma_{0,Y}^2}{n}}\right).}$$

Variances of Both Populations Unknown but Equal

In this case, let $\sigma^2 = \sigma_X^2 = \sigma_Y^2$ be the common (unknown) variance. We need to obtain a pivotal quantity for $\mu_X - \mu_Y$. First observe that, similar to the case of known variances,

$$\frac{(\overline{X}_m - \overline{Y}_n) - (\mu_X - \mu_Y)}{\sqrt{\frac{\sigma^2}{m} + \frac{\sigma^2}{n}}} = \frac{(\overline{X}_m - \overline{Y}_n) - (\mu_X - \mu_Y)}{\sqrt{\sigma^2\left(\frac{1}{m} + \frac{1}{n}\right)}} = Z \sim N(0, 1).$$

Since σ is unknown, we know we have to replace it with a sample SD. The sample SDs of both samples, which may not be equal even though we are assuming the populations SDs are the same, have to be taken into account. We will do that by **pooling** the two SDs. Define

$$\boxed{S_p^2 = \frac{(m - 1)S_X^2 + (n - 1)S_Y^2}{m + n - 2}.}$$

as the **pooled sample variance**. Observe that it is a weighted average of S_X^2 and S_Y^2. We will replace σ by S_p, and we need to find the distribution of

$$\frac{(\overline{X}_m - \overline{Y}_n) - (\mu_X - \mu_Y)}{S_p \sqrt{\frac{1}{m} + \frac{1}{n}}}.$$

We know that

$$\frac{(m-1)S_X^2}{\sigma^2} \sim \chi^2(m-1) \quad \text{and} \quad \frac{(n-1)S_Y^2}{\sigma^2} \sim \chi^2(n-1).$$

But the sum of two independent χ^2 random variables with ν_1 and ν_2 degrees of freedom, respectively, is again a χ^2 random variables with $\nu_1 + \nu_2$ degrees of freedom. Therefore,

$$\frac{(m-1)S_X^2}{\sigma^2} + \frac{(n-1)S_Y^2}{\sigma^2} \sim \chi^2(m+n-2).$$

Again by independence,

$$T = \frac{(\overline{X}_m - \overline{Y}_n) - (\mu_X - \mu_Y)}{\sqrt{\sigma^2 \left(\frac{1}{m} + \frac{1}{n}\right)}} \Bigg/ \sqrt{\frac{(m-1)S_X^2 + (n-1)S_Y^2}{\sigma^2(m+n-2)}} \sim t(m+n-2)$$

since the numerator is distributed as $N(0,1)$ and the denominator is the square root of a χ^2 random variable with $m + n - 2$ degrees of freedom divided by $m + n - 2$. That is, $T \sim t(m + n - 2)$. Using algebra we see that

$$T = \frac{(\overline{X}_m - \overline{Y}_n) - (\mu_X - \mu_Y)}{\sqrt{\sigma^2 \left(\frac{1}{m} + \frac{1}{n}\right)}} \Bigg/ \sqrt{\frac{(m-1)S_X^2 + (n-1)S_Y^2}{\sigma^2(m+n-2)}}$$

$$= \frac{(\overline{X}_m - \overline{Y}_n) - (\mu_X - \mu_Y)}{S_p \sqrt{\frac{1}{m} + \frac{1}{n}}},$$

and so the above expression will be our pivotal quantity for $\mu_X - \mu_Y$. Similar to the derivation in the previous section, we conclude that a $100(1 - \alpha)\%$ observed confidence interval for $\mu_X - \mu_Y$ in the case when both variances are equal but unknown is given by

$$\left((\overline{x}_m - \overline{y}_n) - t(m + n - 2, \alpha/2)s_p \sqrt{\frac{1}{m} + \frac{1}{n}}, \ (\overline{x}_m - \overline{y}_n) + t(m + n - 2, \alpha/2)s_p \sqrt{\frac{1}{m} + \frac{1}{n}} \right).$$

In summary, this is the confidence interval to use when the variances are assumed unknown but equal, and the sample variance we use is the pooled variance because it takes both samples into account.

Variances of Both Populations Unknown and Unequal

The final case occurs if both σ_X^2 and σ_Y^2 are unknown and unequal. Finding a pivotal quantity for $\mu_X - \mu_Y$ with an exact distribution is currently an unsolved problem in statistics called the **Behrens–Fisher** problem. Accordingly, the CI in this case is an approximation and not exact. It turns out it can be shown that

$$T = \frac{(\overline{X} - \overline{Y}_n) - (\mu_X - \mu_Y)}{\sqrt{\frac{S_X^2}{m} + \frac{S_Y^2}{n}}} \sim t(\nu)$$

does follow a t-distribution, but the degrees of freedom is given by the formula

$$\nu = \left\lfloor \frac{\left(\frac{1}{m}r + \frac{1}{n}\right)^2}{\frac{1}{m^2(m-1)}r^2 + \frac{1}{n^2(n-1)}} \right\rfloor,$$

where $r = s_X^2/s_Y^2$ is simply the ratio of the sample variances. Observe that ν depends **only** on the ratio of the sample variances and the sizes of the respective samples, nothing else.

Therefore, an approximate $100(1-\alpha)\%$ observed confidence interval for $\mu_X - \mu_Y$ can now be derived as

$$\left((\overline{x}_m - \overline{y}_n) - t(\nu, \alpha/2)\sqrt{\frac{s_X^2}{m} + \frac{s_Y^2}{n}}, (\overline{x}_m - \overline{y}_n) + t(\nu, \alpha/2)\sqrt{\frac{s_X^2}{m} + \frac{s_Y^2}{n}} \right).$$

This is the CI to use with independent samples from two populations when there is no reason to expect that the variances of the two populations are equal.

Example 4.16   A certain species of beetle is located throughout the United States, but specific characteristics of the beetle, like carapace length, tend to vary by region. An entomologist is studying the carapace length of populations of the beetle located in the southeast and the northeast regions of the country. The data for the two samples of beetles is assumed to be normal and is given below. It is desired to compute a 95% confidence interval for the mean difference in carapace length between the two populations, the variances of which are unknown and assumed to be unequal.

	Carapace Length (in millimeters)			
Northeast	10.5	9.72	10.05	9.94
	8.90	9.33	9.73	11.37
	10.38	10.71	10.38	10.18
	9.93	9.97	10.39	10.86
Southeast	10.51	10.93	10.12	
	10.03	10.54	10.70	
	9.59	10.72		
	9.60	11.21		

Let $N_1, N_2, \ldots, N_{16}$ represent the northeast sample and $S_1, S_2, \ldots, S_{10}$ represent the southeast sample. We first compute the sample variances of the two populations as $s_N^2 = 0.355$ and $s_S^2 = 0.297$. The value of v is given by rounding down

$$\frac{\left(\frac{1}{m}r + \frac{1}{n}\right)^2}{\frac{1}{m^2(m-1)}r^2 + \frac{1}{n^2(n-1)}} = \frac{\left(\frac{1}{16} \cdot \frac{0.355}{0.297} + \frac{1}{10}\right)^2}{\frac{1}{16^2(16-1)}\left(\frac{0.355}{0.297}\right)^2 + \frac{1}{10^2(10-1)}} = 20.579.$$

We find $t(20, 0.025) = \mathrm{invT}(0.975, 20) = 2.0859$. A 95% confidence interval for $\mu_X - \mu_Y$ can now be derived as

$$\left((10.146 - 10.395) - t(20, 0.975)\sqrt{\frac{0.355}{16} + \frac{0.297}{10}},\right.$$
$$\left.(10.146 - 10.395) + t(20, 0.975)\sqrt{\frac{0.355}{16} + \frac{0.297}{10}}\right) = (-0.724, 0.226).$$

Notice that the value 0 is in this confidence interval, and so it is possible that there is no difference in carapace length between the two populations of beetles.

Error Bounds

Error bounds in all the three cases discussed above are easy to derive since the estimator $\overline{X}_m - \overline{Y}_n$ lies in the center of the confidence interval. In particular, for

$$\varepsilon = |(\bar{x}_m - \bar{y}_n) - (\mu_X - \mu_Y)|,$$

we are $100(1 - \alpha)\%$ confident that

$$
\varepsilon \leq
\begin{cases}
z_{\alpha/2} \sqrt{\dfrac{\sigma_{0,X}^2}{m} + \dfrac{\sigma_{0,Y}^2}{n}}, & \text{if both variances are known} \\[3ex]
t(m + n - 2, \alpha/2) s_p \sqrt{\dfrac{1}{m} + \dfrac{1}{n}}, & \text{if both variances are unknown but equal} \\[3ex]
t(v, \alpha/2) \sqrt{\dfrac{s_X^2}{m} + \dfrac{s_Y^2}{n}}, & \text{if both variances are unknown and unequal.}
\end{cases}
$$

4.2.2   RATIO OF TWO NORMAL VARIANCES

Obtaining a confidence interval for the difference of two normal means is the more common practice when comparing two samples. However, sometimes comparing variances is also important. For example, suppose a factory has two machines that produce a part that is used in a product the company makes. The part must satisfy a critical engineering specification of some type, say a dimension specification like diameter, to be usable in the product. It could be that the means of two samples drawn from the machines are not significantly different, but the variances of the two samples are. The parts produced by the machine with the greater variability will deviate from the specification more often than the machine with less variability. Quality control regulations might require that parts that do not conform to the engineering specification be discarded, costing the company money in wasted material, labor, etc. In this situation, it would be meaningful to compare the ratio of the variances of the two machines.

Obtaining a pivotal quantity for σ_X^2 / σ_Y^2 is easy in this case. Recall that

$$
\frac{(m - 1)S_X^2}{\sigma_X^2} \sim \chi^2(m - 1) \quad \text{and} \quad \frac{(n - 1)S_Y^2}{\sigma_Y^2} \sim \chi^2(n - 1).
$$

Since the ratio of two χ^2 random variables each divided by their respective degrees of freedom is distributed as an F random variable, we have

$$
\frac{\frac{(m-1)S_X^2}{\sigma_X^2}}{m - 1} \bigg/ \frac{\frac{(n-1)S_Y^2}{\sigma_Y^2}}{n - 1} = \frac{S_X^2}{\sigma_X^2} \bigg/ \frac{S_Y^2}{\sigma_Y^2} = \frac{S_X^2}{S_Y^2} \frac{\sigma_Y^2}{\sigma_X^2} \sim F(m - 1, n - 1).
$$

It follows that

$$
P\left(F(m - 1, n - 1, 1 - \alpha/2) < \frac{S_X^2}{S_Y^2} \frac{\sigma_Y^2}{\sigma_X^2} < F(m - 1, n - 1, \alpha/2) \right) = 1 - \alpha.
$$

Rewriting so that σ_X^2 / σ_Y^2 is in the center of the inequality gives the probability interval

$$
P\left(\frac{S_X^2}{S_Y^2} \frac{1}{F(m - 1, n - 1, \alpha/2)} < \frac{\sigma_X^2}{\sigma_Y^2} < \frac{S_X^2}{S_Y^2} \frac{1}{F(m - 1, n - 1, 1 - \alpha/2)} \right) = 1 - \alpha.
$$

A $100(1-\alpha)\%$ observed confidence interval for σ_X^2/σ_Y^2 can now be derived as

$$\left(\frac{s_X^2}{s_Y^2}\frac{1}{F(m-1,n-1,\alpha/2)},\frac{s_X^2}{s_Y^2}\frac{1}{F(m-1,n-1,1-\alpha/2)}\right).$$

This confidence interval for σ_X^2/σ_Y^2 can be displayed in a variety of ways. Notice that if $X \sim F(m,n)$, then $1/X \sim F(n,m)$. It follows that $F(n,m,1-\alpha) = 1/F(m,n,\alpha)$. (Prove this!) The confidence interval for σ_X^2/σ_Y^2 can also be expressed as

$$\left(\frac{s_X^2}{s_Y^2}F(n-1,m-1,1-\alpha/2),\frac{s_X^2}{s_Y^2}F(n-1,m-1,\alpha/2)\right)$$

$$=\left(\frac{s_X^2}{s_Y^2}\frac{1}{F(m-1,n-1,\alpha/2)},\frac{s_X^2}{s_Y^2}F(n-1,m-1,\alpha/2)\right)$$

$$=\left(\frac{s_X^2}{s_Y^2}F(n-1,m-1,1-\alpha/2),\frac{s_X^2}{s_Y^2}\frac{1}{F(m-1,n-1,1-\alpha/2)}\right).$$

Example: Continuing Example 4.16 involving beetle populations, set $\alpha = 0.025$. At the 95% confidence level, $F(15,9,0.025) = 3.7694$ and $F(15,9,0.975) = 0.3202$. Therefore, a 95% confidence interval for σ_N^2/σ_S^2 is given by

$$\left(\frac{0.355}{0.297}\cdot\frac{1}{3.7694},\frac{0.355}{0.297}\cdot\frac{1}{0.3202}\right) = (0.317\,1, 3.\,732\,9).$$

Since the interval contains the value 1, we cannot conclude the variances are different at the 95% confidence level.

We summarize, on Table 4.3, all the two-sample confidence intervals obtained for the normal distribution.

4.2.3   DIFFERENCE OF TWO BINOMIAL PROPORTIONS

Let $X \sim \text{Binom}(m, p_X)$ and $Y \sim \text{Binom}(n, p_Y)$ be two binomial random variables with parameters p_X and m, and p_Y and n, respectively. From the CLT, we know that for sufficiently large sample sizes, the distributions of $\overline{X}_m$ and $\overline{Y}_n$ can be approximated as

$$\overline{X}_m \sim N\left(p_X, \sqrt{\frac{p_X(1-p_X)}{m}}\right) \text{ and } \overline{Y}_n \sim N\left(p_Y, \sqrt{\frac{p_Y(1-p_Y)}{n}}\right).$$

By independence of the samples, the distribution of $\overline{X}_m - \overline{Y}_n$ can be approximated as

$$\overline{X}_m - \overline{Y}_n \sim N\left(p_X - p_Y, \sqrt{\frac{p_X(1-p_X)}{m} + \frac{p_Y(1-p_Y)}{n}}\right).$$

Table 4.3: Confidence intervals for the parameters of the normal distribution (two samples)

Parameter	Conditions	Confidence Interval
$d = \mu_X - \mu_Y$	$\sigma_X^2 = \sigma_{0,X}^2$ and $\sigma_Y^2 = \sigma_{0,Y}^2$ known	$\left((\bar{x}_m - \bar{y}_n) - z_{\alpha/2} \sqrt{\frac{\sigma_{0,X}^2}{m} + \frac{\sigma_{0,Y}^2}{n}}, \right.$ $\left. (\bar{x}_m - \bar{y}_n) + z_{\alpha/2} \sqrt{\frac{\sigma_{0,X}^2}{m} + \frac{\sigma_{0,Y}^2}{n}} \right)$
$d = \mu_X - \mu_Y$	σ_X^2, σ_Y^2 unknown, $\sigma_X^2 = \sigma_Y^2$	$\left((\bar{x}_m - \bar{y}_n) - t(m+n-2, \alpha/2) s_p \sqrt{\frac{1}{m} + \frac{1}{n}}, \right.$ $\left. (\bar{x}_m - \bar{y}_n) + t(m+n-2, \alpha/2) s_p \sqrt{\frac{1}{m} + \frac{1}{n}} \right)$
$d = \mu_X - \mu_Y$	σ_X^2, σ_Y^2 unknown, $\sigma_X^2 \neq \sigma_Y^2$	$\left((\bar{x}_m - \bar{y}_n) - t(\nu, \alpha/2) \sqrt{\frac{s_X^2}{m} + \frac{s_Y^2}{n}}, \right.$ $\left. (\bar{x}_m - \bar{y}_n) + t(\nu, \alpha/2) \sqrt{\frac{s_X^2}{m} + \frac{s_Y^2}{n}} \right),$ $\nu = \left\lfloor \frac{\left(\frac{1}{m} r + \frac{1}{n} \right)^2}{\frac{1}{m^2(m-1)} r^2 + \frac{1}{n^2(n-1)}} \right\rfloor, r = \frac{s_X^2}{s_Y^2}$
$r = \frac{\sigma_X^2}{\sigma_Y^2}$	μ_X, μ_Y unknown	$\left(\frac{s_X^2}{s_Y^2} \frac{1}{F(m-1, n-1, \alpha/2)}, \frac{s_X^2}{s_Y^2} \frac{1}{F(m-1, n-1, 1-\alpha/2)} \right)$

Therefore,

$$Z = \frac{(\overline{X}_m - \overline{Y}_n) - (p_X - p_Y)}{\sqrt{\frac{p_X(1-p_X)}{m} + \frac{p_Y(1-p_Y)}{n}}} \sim N(0,1) \text{ (approximate).}$$

Approximating p_X by $\bar{p}_X = \bar{x}_m$ and p_Y by $\bar{p}_Y = \bar{y}_n$, we obtain the $100(1-\alpha)\%$ (approximate) confidence interval for $p_X - p_Y$ as

$$\left((\bar{p}_X - \bar{p}_Y) - z_{\alpha/2} \sqrt{\frac{\bar{p}_X(1-\bar{p}_X)}{m} + \frac{\bar{p}_Y(1-\bar{p}_Y)}{n}}, (\bar{p}_X - \bar{p}_Y) + z_{\alpha/2} \sqrt{\frac{\bar{p}_X(1-\bar{p}_X)}{m} + \frac{\bar{p}_Y(1-\bar{p}_Y)}{n}} \right).$$

The error $\varepsilon = |(\bar{p}_X - \bar{p}_Y) - (p_X - p_Y)|$ can be bounded (approximately) as

$$\varepsilon \leq z_{\alpha/2} \sqrt{\frac{\bar{p}_X(1-\bar{p}_X)}{m} + \frac{\bar{p}_Y(1-\bar{p}_Y)}{n}}$$

at a $100(1-\alpha)\%$ confidence level.

Example 4.17   An experiment was conducted by a social science researcher which involved assessing the benefits of after-school enrichment programs in a certain low-income school district in Cleveland, Ohio. A total of 147 three- and four-year-old children were involved in the study. Children were randomly assigned to two groups, one group of 73 students which participated in the after-school programs, and a control group with 74 students which did not participate in these programs. The children were followed as adults, and a number of data items were collected, one of them being their annual incomes. In the control group, 23 out of the 74 children were earning more that $75,000 per year whereas in the group participating in the after-school programs, 38 out of the 73 children were earning more than $75,000 per year. Let p_X and p_Y denote the proportion of three- and four-year-old children who do and do not participate in after-school programs making more than $75,000 per year, respectively. A 95% confidence interval for $p_X - p_Y$ is given by

$$(0.520 - 0.311) \pm (1.96) \sqrt{\frac{0.520(1 - 0.520)}{73} + \frac{0.311(1 - 0.311)}{74}} \implies (0.053, 0.365).$$

As a result of the study, we can be 95% confident that between 5.3% and 36.5% more children not having participated in after-school enrichment programs were earning less than $75,000 per year than students who did participate in these programs.

4.2.4   PAIRED SAMPLES

In many situations we wish to compare the means or proportions of two populations, but we cannot assume that the random samples from each population are independent. For example, if we want to compare exam scores before and after a learning module is taken, the samples are not independent because they are the exam scores of the same students before and after the learning module. Another example would be a weight loss or drug effectiveness program. Clearly, the random sample must involve the same people, tested before the program and after the program.

Let $X_1, X_2, \ldots, X_m$ and $Y_1, Y_2, \ldots, Y_m$ be random samples of the **same size m** from two populations where X_i is paired with Y_i. The variables X and Y from which the corresponding samples are drawn are **dependent** which means the procedure for independent samples cannot be used. The solution is to consider the random variable for the difference between X and Y,

$$D = X - Y \text{ and } D_i = X_i - Y_i, \quad i = 1, 2, \ldots, m.$$

Clearly, $D_1, D_2, \ldots, D_m$ is a random sample from D, the population of differences, that has mean $\mu_D = \mu_X - \mu_Y$. **We will assume that** $D \sim N(\mu_D, \sigma_D)$. This is now just like a one-sample CI with an unknown variance. Since the variance σ_D^2 is unknown, it must be estimated from the sample of differences as

$$S_D^2 = \frac{1}{m-1} \sum_{i=1}^{m} \left(D_i - \overline{D}_m\right)^2.$$

Therefore, $\dfrac{\overline{D}_m - \mu_D}{S_D/\sqrt{m}} = T \sim t(m - 1)$. A $100(1 - \alpha)\%$ confidence interval for the difference between the two means μ_D can now be obtained as

$$\left(\overline{d}_m - t(m-1,\alpha/2)\frac{s_D}{\sqrt{m}}, \overline{d}_m + t(m-1,\alpha/2)\frac{s_D}{\sqrt{m}}\right).$$

As usual, d_i is the observed difference of the ith pair $x_i - y_i$.

Example 4.18 Ten middle-aged men with high blood pressure engage in a regimen of aerobic exercise on a newly introduced type of tread mill. Their blood pressures are recorded before starting the exercise program. After using the treadmill for 30 minutes each day for six months, their blood pressures are again recorded. During the six-month period, the ten men do not change their lifestyles in any other significant way. The two blood pressure readings in mmHg (before and after the period of aerobic exercise) are given in the table below.

Male i	BP (before)(X_i)	BP (after) (Y_i)	Difference D_i
1	143	144	-1
2	171	164	7
3	160	149	11
4	182	175	7
5	149	142	7
6	162	162	0
7	177	173	4
8	165	156	9
9	150	148	2
10	165	161	4

Take $\alpha = 0.05$. From the table, we compute $\overline{d}_{10} = 5$ and $s_D = 3.887$. A 95% confidence interval for μ_D is given by

$$\left(5 - t(9, 0.025)\frac{3.887}{\sqrt{10}}, 5 + t(9, 0.025)\frac{3.887}{\sqrt{10}}\right) = (2.22, 7.78).$$

We are 95% confident that exercising on the tread mill can lower blood pressure roughly between 2 and 8 points.

4.3 PREDICTION INTERVALS

Confidence intervals are used to estimate unknown parameters such as μ or σ in a normal distribution, or p in the binomial distribution. Prediction intervals are used not to estimate parameters but rather to estimate **future** sample values drawn from the distribution. The parameters in the population such as μ or σ may be unknown.

Suppose $X_1, X_2, \ldots, X_n$ is a random sample from a random variable X. Consider making another observation from X, namely X_{n+1}. Given a value of α, we would like to derive an interval with random endpoints (as in the confidence interval case) such that the probability of the $(n + 1)st$ observation X_{n+1} lying in the interval, based on the sample $X_1, \ldots, X_n$, is $1 - \alpha$.

Definition 4.19   Let $X_1, X_2, \ldots, X_n$ be a random sample from a random variable X. A $100(1 - \alpha)\%$ **prediction interval** for X_{n+1} is an interval (having random endpoints) of the form $l(X_1, X_2, \ldots, X_n)$ and $u(X_1, X_2, \ldots, X_n)$ such that

$$P \left(l \left(X_1, X_2, \ldots, X_n \right) < X_{n+1} < u \left(X_1, X_2, \ldots, X_n \right) \right) = 1 - \alpha.$$

To obtain the actual prediction interval, we substitute the observed values of the sample into the functions $l(X_1, X_2, \ldots, X_n)$ and $u(X_1, X_2, \ldots, X_n)$ to get an interval $(l(x_1, \ldots, x_n), u(x_1, \ldots, x_n))$ of real numbers. Unlike for confidence intervals, the statement $P(l(x_1, \ldots, x_n) < X_{n+1} < u(x_1, \ldots, x_n))$ does have meaning since X_{n+1} is a random variable. We are not capturing a number in the interval, but a random variable.

Similar to confidence intervals, constructing prediction intervals depends on identifying **pivotal** quantities. In a prediction interval, a pivotal quantity may depend on $X_1, \ldots, X_{n+1}$.

Definition 4.20   Let $X_1, X_2, \ldots, X_n$ be a random sample from a random variable X. A pivotal quantity for X_{n+1} is a random variable $h(X_1, X_2, \ldots, X_n, X_{n+1})$ whose distribution is independent of the sample values. **It is assumed that the underlying population follows a normal distribution**.

We will obtain pivotal quantities and construct prediction intervals depending on whether μ or σ is known or unknown.

$\mu = \mu_0$ and $\sigma^2 = \sigma_0^2$ are Known

This case is straightforward since no parameters have to be estimated. We only want to predict the next sample value from the previous sample values. Since μ_0 and σ_0 are known, $\frac{X_{n+1} - \mu_0}{\sigma_0} = Z \sim N(0, 1)$, and therefore

$$P \left(-z_{\alpha/2} < \frac{X_{n+1} - \mu_0}{\sigma_0} < z_{\alpha/2} \right) = P \left(\mu_0 - z_{\alpha/2}\sigma_0 < X_{n+1} < \mu_0 + z_{\alpha/2}\sigma_0 \right) = 1 - \alpha.$$

Therefore, a $100(1 - \alpha)\%$ prediction interval for X_{n+1} is simply

$$\boxed{\left(\mu_0 - z_{\alpha/2}\sigma_0, \mu_0 - z_{\alpha/2}\sigma_0 \right).}$$

In other words, we can predict a sample value from $N(\mu_0, \sigma_0)$ will be in $(\mu_0 - z_{\alpha/2}\sigma_0, \mu_0 + z_{\alpha/2}\sigma_0)$ with probability $1 - \alpha$.

μ **is Unknown and** $\sigma^2 = \sigma_0^2$ **is Known**

Since $Var(X_{n+1} - \overline{X}) = \sigma_0^2 + \sigma_0^2/n$, by independence, we have

$$\frac{X_{n+1} - \overline{X}}{\sqrt{\sigma_0^2 + \frac{\sigma_0^2}{n}}} = \frac{X_{n+1} - \overline{X}}{\sigma_0\sqrt{1 + \frac{1}{n}}} = Z \sim N(0, 1).$$

Therefore, the random variable $\dfrac{X_{n+1} - \overline{X}}{\sigma_0\sqrt{1 + \frac{1}{n}}}$ is a pivotal quantity for X_{n+1}. The $100(1 - \alpha)\%$

prediction interval for X_{n+1} is given by

$$\left(\overline{x} - z_{\alpha/2}\sigma_0\sqrt{1 + \frac{1}{n}}, \ \overline{x} + z_{\alpha/2}\sigma_0\sqrt{1 + \frac{1}{n}} \right).$$

$\mu = \mu_0$ **is Known and** σ^2 **is Unknown**

In this case,

$$\frac{X_{n+1} - \mu_0}{S_X} = T \sim t(n - 1),$$

and so the $100(1 - \alpha)\%$ prediction interval for X_{n+1} is now given by

$$\left(\mu_0 - t(n - 1, \alpha/2)s_X, \ \mu_0 + t(n - 1, \alpha/2)s_X \right).$$

Both μ **and** σ^2 **are Unknown**

In this case,

$$\frac{X_{n+1} - \overline{X}}{\sqrt{S_X^2 + \frac{S_X^2}{n}}} = \frac{X_{n+1} - \overline{X}}{S_X\sqrt{1 + \frac{1}{n}}} = T \sim t(n - 1).$$

The $100(1 - \alpha)\%$ prediction interval for X_{n+1} is now given by

$$\left(\overline{x} - t(n - 1, \alpha/2)s_X\sqrt{1 + \frac{1}{n}}, \ \overline{x} + t(n - 1, \alpha/2)s_X\sqrt{1 + \frac{1}{n}} \right).$$

One of the uses of prediction intervals is in the detection of **outliers**, extreme values of the random variable or a value that comes from a population whose mean is different from the one under consideration. **Given a choice of** α**, the sample value** X_{n+1} **will be considered an outlier if** X_{n+1} **is not in the** $100(1 - \alpha)\%$ **prediction interval for** X_{n+1}**.**

Example 4.21   An online furniture retailer has collected the times it takes an adult to assemble a certain piece of its furniture. The data in the table below represents a random sample $X_1, X_2, \ldots, X_{36}$ of the number of minutes it took 36 adults to assemble a certain advertised "easy to assemble" outdoor picnic table. Assume the data is drawn from a normal population.

Assembly Time for Adults											
17	13	18	19	17	21	29	22	16	28	21	15
26	23	24	20	8	17	17	21	32	18	25	22
16	10	20	22	19	14	30	22	12	24	28	11

We will use the data in the table to construct a 95% prediction interval for the next assembly time. We compute the mean and standard deviation of the sample as $\bar{x}_{36} = 19.92$ and $s_X = 5.73$. The 95% prediction interval is given by

$$\left(19.92 - t(35, 0.025)\,(5.73)\,\sqrt{1 + \frac{1}{36}}, \, 19.92 + t(35, 0.025)\,(5.73)\,\sqrt{1 + \frac{1}{36}} \right)$$

$$= (8.13, 31.71).$$

We can write a valid probability statement as $P(8.13 < X_{37} < 31.71) = 0.95$.

Any data value falling outside the prediction interval $(8.13, 31.71)$ could be considered an outlier at the 95% level.

4.4   PROBLEMS

4.1.   An urn contains only black and white marbles with unknown proportions. If a random sample of size 100 is drawn and it contains 47 black marbles, find the following.

   (a) The percentage of black marbles in the urn is estimated as _____with a standard error of _____

   (b) The SE measures the likely size of the error due to chance in the estimate of the percentage of black marbles in the urn. (T/F)

   (c) Suppose your estimate of the proportion of black marbles in the urn is $\hat{p}$. Then $\hat{p}$ is likely to be off from the true proportion of black marbles in the urn by the SE. (T/F)

   (d) A 95% CI for the proportion of black marbles in the urn is _____to

   _____

   (e) What is a 95% CI for the proportion of black marbles in the sample, or does that make sense? Explain.

   (f) Suppose we know that 53% of the marbles in the urn are black. We take a random sample of 100 marbles and calculate the SE as 0.016. Find the chance that the proportion of black marbles in the sample is between $0.53 - 0.032$ and $0.53 + 0.032$.

4.2.   Suppose a random sample of 100 male students is taken from a university with 546 male students in order to estimate the population mean height. The sample mean is $\bar{x} = 67.45$ inches with an SD of 2.93 inches.

(a) Assuming the sampling is done with replacement, find a 90% CI for the population mean height.

(b) Assuming the sampling is done without replacement, find a 90% CI.

4.3. Suppose 50 random samples of size 10 are drawn from a population which is normally distributed with mean 40 and variance 3. A 95% confidence interval is calculated for each sample.

(a) How many of these 50 CIs do you expect to contain the true population mean $\mu = 40$?

(b) If we define the rv X to be the number of intervals out of 50 which contain the true mean $\mu = 40$, what is the distribution of X? Find $P(X = 40)$, $P(X \le 40)$, and $P(X > 45)$.

4.4. Find a conservative estimate of the sample size needed in order to ensure the error in a poll is less than 3%. Assume we are using 95% confidence.

4.5. Find the 99% confidence limits for the population proportion of voters who favor a candidate. The sample size is 100, and the sample percentage of voters favoring the candidate was 55%.

4.6. A random sample is taken from a population which is $N(\mu, 5)$. The sample size is 20 and results in the sample mean $\bar{x} = 15.2$.

(a) Find the CI for levels of confidence 70, 80, and 90%.

(b) Repeat the problem assuming the sample size is 100.

4.7. Show that a lower $100(1 - \alpha)\%$ one-sided confidence interval for the unknown mean μ with unknown variance is given by $(\bar{x} - t(n - 1, \alpha)s_X/\sqrt{n}, +\infty)$.

4.8. Suppose $X_1, \ldots, X_n$ is a random sample from a continuous random variable X with population median m. Suppose that we use the interval with random endpoints $(X_{\min}, X_{\max})$ as a CI for m.

(a) Find $P(X_{\min} < m < X_{\max})$ giving the confidence level for the interval $(X_{\min}, X_{\max})$. Notice that it is not 100% because this involves a random sample.

(b) Find $P(X_{\min} < m < X_{\max})$ if the sample size is $n = 8$.

4.9. A random sample of 225 flights shows that the mean number of unoccupied seats is 11.6 with SD $= 4.1$. Assume this is the population SD.

(a) Find a 90% CI for the population mean.

(b) The 90% CI you found means (choose ONE):

   (a) The interval contains the population mean with probability 0.9.

   (b) If repeated samples are taken, 90% of the CIs contain the population mean.

   (c) What minimum sample size do you need if you want the error reduced to 0.2?

4.10. Suppose you want to provide an accurate estimate of customers preferring one brand of coffee over another. You need to construct a 95% CI for p so that the error is at most 0.015. You are told that preliminary data shows $p = 0.35$. What sample size should you choose?

4.11. Consider the following data points for fuel mileage in a particular vehicle.

$$
\begin{array}{cccccccccc}
42 & 36 & 38 & 45 & 41 & 47 & 33 & 38 & 37 & 36 \\
40 & 44 & 35 & 39 & 38 & 41 & 44 & 37 & 37 & 49
\end{array}
$$

Assume these data points are from a random sample. Construct a 95% CI for the mean population mpg. Is there any reason to suspect that the data does not come from a normal population? What would a lower one-sided CI be, and how would it be interpreted?

4.12. The accuracy of speedometers is checked to see if the SD is about 2 mph. Suppose a random sample of 35 speedometers are checked, and the sample SD is 1.2 mph. Construct a 95% CI for the population variance.

4.13. The ACT exam for an entering class of 535 students had a mean of 24 with an SD of 3.9. Assume this class is representative of future students at this college.

   (a) Find a 90% CI for the mean ACT score of all future students.

   (b) Find a 90% for the ACT score of a future student.

4.14. A random sample of size 32 is taken to assess the weight loss on a low-fat diet. The mean weight loss is 19.3 pounds with an SD of 10.8 pounds. An independent random sample of 32 people who are on a low calorie diet resulted in a mean weight loss of 15.1 pounds with an SD of 12.8 pounds. Construct a 95% CI for the mean difference between a low calorie and a low fat diet. Do not pool the data.

4.15. A sample of 140 LEDs resulted in a mean lifetime of 9.7 years with an SD of 6 months. A sample of 200 CFLs resulted in a mean lifetime of 7.8 years with an SD of 1.2 years. Find a 95% and 99% CI for the difference in the mean lifetimes. Is there evidence that the difference is real? Do not pool the data.

4.16. The SD of a random sample of the service times of 200 customers was found to be 3 minutes.

   (a) Find a 95% CI for the SD of the service times of all such customers.

(b) How large a sample is needed in order to be 99.93% confident that the true population SD will not differ from the sample SD by more than 5%.

4.17. Suppose the sample mean number of sick days at a factory is 6.3 with a sample SD of 4.5. This is based on a sample of 25.

(a) Find a 98% CI for the population mean number of sick days.

(b) Calculate the sample size needed so that a 95% CI has an error of no more than 0.5 days.

4.18. A random sample of 150 colleges and universities resulted in a sample mean ACT score of 20.8 with an SD of 4.2. Assuming this is representative of all future students,

(a) find a 95% CI for the mean ACT of all future students, and

(b) find a 95% for the ACT score of a single future student.

4.19. A logic test is given to a random sample of students before and after they completed a formal logic course. The results are given below. Construct a 95% confidence interval for the mean difference between the before and after scores.

After	74	83	75	88	84	63	93	84	91	77
Before	73	77	70	77	74	67	95	83	84	75

4.20. Two independent groups, chosen with random assignments, A and B, consist of 100 people each of whom have a disease. An experimental drug is given to group A but not to group B, which are termed treatment and control groups, respectively. Two simple random samples have yielded that in the two groups, 75 and 65 people, respectively, recover from the disease. To study the effect of the drug, build a 95% confidence interval for the difference in proportions $p_A - p_B$.

CHAPTER 5

Hypothesis Testing

This chapter is one of the cornerstones of statistics because it allows us to reach a decision based on an experiment with random outcomes. The basic question in an experiment is whether or not the outcome is **real**, or simply **due to chance variation**. For example, if we flip a coin 100 times and obtain 57 heads, can we conclude the coin is not fair? We know we expect 50 heads, so is 57 too many, or is it due to chance variation? Hypothesis testing allows us to answer such questions. This is an extremely important issue, for instance in drug trials in which we need to know if a drug truly is efficacious, or if the result of the clinical trial is simply due to chance, i.e., the possibility that a subject will simply improve or get worse on their own. The purpose of this chapter is to show how hypothesis testing is implemented.

5.1   A MOTIVATING EXAMPLE

In 2004, the drug company Merck was conducting a clinical trial[1] to determine if its drug VIOXX© had any effect in treating polyps in the colon. VIOXX© is a drug originally designed to treat chronic inflammation for arthritis patients so this would have been an added benefit of the drug if the clinical trial turned out positively. The trial involved 2,586 patients in a controlled, randomized, double-blind experiment. In the experiment, 1,299 patients were assigned to the control group and 1,287 to the treatment group. Among other things, the safety of the drug was of paramount importance, and the clinical trial monitored the drug for safety. At the conclusion of the trial, it was observed that 26 of the control group and 46 of the treatment group experienced a cardiovascular event (CV). Can this be attributed to the drug? Or is it due to random chance?

The first part of that question is answered by the way the experiment was designed. By making the control and treatment groups statistically identical except that one group took the drug and the other did not, we can be reasonably sure that if the difference in CV events is real, then we can attribute it to the drug. But how do we know the difference is real? Couldn't it be due to—simply by chance—assigning more people who were destined to have a CV event to the treatment group rather than to the control group? To answer that question, because we randomly assigned subjects to the treatment and control groups, we may apply a probabilistic method to quantify this. That method is hypothesis testing.

---

[1]Cardiovascular events associated with rofecoxib in a colorectal adenoma chemoreception trial, R.S. Bristlier et al., March 17, 2005, *N. Engl. J. Med.*, 2005; 352:1092–1102.

What follows is how hypothesis testing would work in this situation. We will introduce the necessary terminology as we explain the method.

First, let's denote the true population proportion of people who will have a CV event as p_T if they are in the treatment group, and p_C if they are in the control group. Hypothesis testing assumes that the difference between these two quantities should be zero, i.e., that there is no difference between the two groups. We say that this is the **Null Hypothesis**, and write it as $H_0 :$ $p_T - p_C = 0$. Next, since we **observed** $\overline{p}_T = 0.0357$ and $\overline{p}_C = 0.0200$, the observed difference $\overline{p}_T - \overline{p}_C = 0.0157 > 0$ and so establishes the **Alternative Hypothesis** as $H_1 : p_T - p_C > 0$.

What are the chances of observing a difference of 0.0157 in the proportions, if, in fact the difference should be 0 **under the assumption of the null hypothesis**? The random assignment of subjects to control and treatment groups allows us to use probability to actually answer this question. We will see by the methods developed in this chapter that we will be able to calculate that the chance of observing a difference of 0.0157 (or more) is 0.00753, under the assumption of the null hypothesis that there should be no difference. This value is called the **p-value of the test** or the **level of significance of the test**.

Now we are ready to reach a conclusion. Under the assumption there is no difference, we calculate that the chance of observing a difference of 0.0157 is only 0.7%. But, this is what actually occurred even though it is extremely unlikely. The more likely conclusion, supported by the evidence, is that our assumption is incorrect, i.e., it is more likely H_0 is wrong.[2] The conclusion is that we reject the null hypothesis in favor of the alternative hypothesis. How small does the p-value have to be for a conclusion to reject the null? Statisticians use the guide in Table 5.1 in Section 5.3.2 to reach a conclusion based on the p-value.

Remark 5.1   What does it mean to **reject the null?** It means that our assumption that the null is true leads to a very low probability of it actually being true. So either we just witnessed a very low probability event, or the null is not true. We say that the null is rejected if the p-value of the test is low enough. If the p-value is not low enough, it means that it is **plausible** the null is true, i.e., there is not enough evidence against it, and therefore we do not reject it.

What we just described is called the **p-value approach** to hypothesis testing, and the p-value is essentially the probability we reach a wrong conclusion if we assume the null. Another approach is called the **critical value approach**. We'll discuss this in more detail later, but in this example here's how it works.

First, we specify a **level of significance** α, which is the largest p-value we are willing to accept to reject the null. Say we take $\alpha = 0.01$. Then, we calculate the z-value that gives the proportion α to the right of z under the standard normal curve, i.e., the 99th percentile of the normal curve. In this case $z = 2.326$. We are choosing the area to the right because our alternative hypothesis is $H_1 : p_T - p_C > 0$. This z-value is called the **critical value** for a 1% level of significance. Now, when we carry out the experiment, any value of the test statistic

---

[2]In 2004, on the basis of statistical evidence, Merck pulled VIOXX© off the market.

$z = \dfrac{\text{observed} - \text{expected}}{SE} \geq 2.326$ will result in our rejecting the null, and we know the corresponding p-value must be less than $\alpha = 0.01$. The advantage of this approach is that we know the value of the test statistic we need to reject the null before we carry out the experiment.

Example 5.2   We have a coin which we think is not fair. That means we think $p \neq 0.5$, where p is the probability of flipping a head. Suppose we toss the coin 100 times and get 60 heads. The sample proportion of heads we obtain is $\overline{p} = 0.6$. Thus, our hypothesis test is $H_0 : p = 0.5$ vs. $H_1 : p > 0.5$ because we got 60% heads. We start off assuming it is a fair coin.

Now suppose we are given a level of significance $\alpha = 0.05$. We know, using the normal approximation, that the sample proportion $\overline{P} \sim N(0.5, \sqrt{0.5 \times 0.5/100})$. The critical value for $\alpha = 0.05$ is the z-value giving 0.05 area to the right of z, and this is $z = 1.644$. This is the critical value for this test. The value of the test statistic is $z = \frac{0.6-0.5}{0.05} = 2.0$. Consequently, since $2.0 > 1.644$, we conclude that we may reject the null at the 5% level of significance.

The p-value approach gives more information since it actually tells us the chance of being wrong assuming the null. In this case the p-value is $P(\overline{P} \geq 0.6) = \text{normalcdf}(.6, \infty, .5, .05) = 0.0227$, or by standardizing,

$$P(\overline{P} \geq 0.6 \mid p = 0.5) = P\left(\frac{\overline{P} - 0.5}{\sqrt{0.5 \times 0.5/100}} \geq \frac{0.6 - 0.5}{\sqrt{0.5 \times 0.5/100}}\right)$$

$$\approx P(Z \geq 2.0) = 0.0227.$$

We have calculated the chance of getting 60 or more heads if it is a fair coin as only 2%, so we have significant evidence against the null. Again, since $0.02 < 0.05$, we reject the null and conclude it is not a fair coin. We could be wrong, but the chance of being wrong is only 2%. Another important point is our choice of null $H_0 : p = 0.5$ which specifies exactly what p should be for this coin if it's fair.

5.2   THE BASICS OF HYPOTHESIS TESTING

To start with, suppose that $X_1, X_2, \ldots, X_n$ is a random sample from a random variable X with pdf $f_X(x; \vartheta)$. We would like to devise a test to decide whether we should reject the value $\vartheta = \vartheta_0$ in favor of some other value of ϑ. Consider the two hypotheses

$$H_0 : \vartheta = \vartheta_0 \quad \text{vs.} \quad H_1 : \vartheta \neq \vartheta_0. \tag{5.1}$$

The hypothesis $H_0 : \vartheta = \vartheta_0$ is called the **null** hypothesis. The hypothesis $H_1 : \vartheta \neq \vartheta_0$ is called the **two-sided alternative** hypothesis.

The Critical Value Approach and the Connection with CIs

Let $(l(X_1, X_2, \ldots, X_n), u(X_1, X_2, \ldots, X_n))$ be a $100(1 - \alpha)\%$ confidence interval for ϑ. We are $100(1 - \alpha)\%$ confident that (l, u) contains the true value of ϑ, and so if we have a result from a random sample that is not in this interval, that is sufficiently good evidence our assumption about the value of ϑ may not be a good one. Accordingly, we will **reject** H_0 if

$$\vartheta_0 \leq l(X_1, X_2, \ldots, X_n), (\vartheta_0 \text{ appears to be too small})$$

or if

$$\vartheta_0 \geq u(X_1, X_2, \ldots, X_n), (\vartheta_0 \text{ appears to be too large}).$$

As before,

$$P(\vartheta_0 \leq l(X_1, X_2, \ldots, X_n) \text{ or } \vartheta_0 \geq u(X_1, X_2, \ldots, X_n) \mid \vartheta = \vartheta_0) = \alpha.$$

That is, the probability of rejecting the null hypothesis when the null is true ($\vartheta = \vartheta_0$) is α. The form of the pivotal quantity for the CI under the null hypothesis is called the **test statistic**.

The value of α is called the **significance level** (or simply **level**) of the test, and this is the largest probability we are willing to accept for making an error in rejecting a true null. Rejecting the null hypothesis when the null is true is commonly called a **Type I** error. A **Type II** error occurs when we do not reject the null hypothesis when it is false. The probability of making a Type II error is denoted β. To summarize,

$$\boxed{\alpha = P(\text{reject } H_0 \mid H_0 \text{ is true}) \quad \text{and} \quad \beta = P(\text{fail to reject } H_0 \mid H_0 \text{ is false}).}$$

The table may make this easier to remember.

		Conclusion	
		Retain H_0	Reject H_0
Actually	H_0 true	Correct	Type I error; probability α
	H_0 false	Type II error; probability β	Correct

It turns out that the value of β is not fixed like the value of α and depends on the value of ϑ taken in the alternative hypothesis $H_1 : \vartheta = \vartheta_1 \neq \vartheta_0$. In general, the value of β is different for every choice of $\vartheta_1 \neq \vartheta_0$, and so $\beta = \beta(\vartheta_1)$ is a function of ϑ_1.

Remark 5.3   A Type I error rejects a true null, while a Type II error does not reject a false null. In general, making α small is of primary importance. Here's an analogy to explain why this is. Suppose a male suspect has been arrested for murder. He is guilty or not guilty. In the U.S., the null hypothesis is H_0 : **suspect is not guilty**, with alternative H_1 : **suspect is guilty**. A Type I error, measured by α, says that the suspect is found guilty by a jury, but he is really not guilty. A

Type II error, measured by β, says that the suspect is found not guilty by the jury, when, in fact, he is guilty. Both of these are errors, but finding an innocent man guilty is considered the more serious error. That's one of the reasons we focus on keeping α small.

The set of hypotheses in (5.1) is called a **two-sided** test. In such a test, we are interested in deciding if $\vartheta = \vartheta_0$ or not. Sometimes we are only interested in testing whether $H_0 : \vartheta = \vartheta_0$ vs. $H_1 : \vartheta < \vartheta_0$ or vs. $H_1 : \vartheta > \vartheta_0$. These types of tests are called **one-sided**. Often one decides the form of the alternative on the basis of the result of the experiment. For example, in our coin tossing experiment we obtained $\overline{p} = 0.6$, so that a reasonable alternative is $H_1 : p > 0.5$.

To construct a test of hypotheses for one-sided tests using the critical value approach, we simply use **one-sided confidence intervals** instead of two-sided. Specifically, for a given level of significance α, for the alternative hypothesis $H_1 : \vartheta < \vartheta_0$, we will reject $H_0 : \vartheta = \vartheta_0$ in favor of H_1 if ϑ_0 is not in the $100(1-\alpha)\%$ confidence interval $(-\infty, u(X_1, X_2, \ldots, X_n))$ for ϑ. In particular, we reject the null hypothesis if $\vartheta_0 \geq u(x_1, \ldots, x_n)$ if $X_1 = x_1, X_2 = x_2, \ldots, X_n = x_n$ are the observed sample values. For the alternative hypothesis, $H_1 : \vartheta > \vartheta_0$, we reject H_0 if $\vartheta_0 \leq l(x_1, \ldots, x_n)$. Finally, the **the set of real number S for which we reject H_0 if $\vartheta_0 \in S$** is called the **critical region** of the test. To summarize, we reject the null hypothesis if

$$\vartheta_0 \geq u(x_1, \ldots, x_n), \qquad\qquad \text{for } H_1 : \vartheta < \vartheta_0$$
$$\vartheta_0 \leq l(x_1, \ldots, x_n), \qquad\qquad \text{for } H_1 : \vartheta > \vartheta_0$$
$$\vartheta_0 \geq u(x_1, \ldots, x_n) \text{ or } \vartheta_0 \leq l(x_1, \ldots, x_n) \qquad \text{for } H_1 : \vartheta \neq \vartheta_0.$$

5.3   HYPOTHESES TESTS FOR ONE PARAMETER

In this section we develop tests of hypotheses (both one- and two-sided) for **one** unknown parameter. We will focus on the normal parameters μ and σ, and on the binomial proportion p.

All the hypotheses tests for means or proportions are based on a test statistic of the form $\dfrac{\text{observed-expected}}{SE}$. This always involves calculating the correct SE. In hypothesis testing, the **expected** in the numerator is always calculated on the basis of the null hypothesis.

5.3.1   HYPOTHESES TESTS FOR THE NORMAL PARAMETERS, CRITICAL VALUE APPROACH

Tests of hypotheses for the parameters μ and σ of a random variable $X \sim N(\mu, \sigma)$ can be easily derived from the confidence intervals described in Chapter 4. The decision to either reject or not reject the null hypothesis is given in terms of critical values and the values of the respective test statistics for a given random sample from X. The value of a statistic when the data values from a random sample are substituted for the random variables in the statistic is called a **score**.

Test for μ, $\sigma = \sigma_0$ Known, Null $H_0 : \mu = \mu_0$

The SE is $\sigma_0/\sqrt{n}$. The test statistic is $Z = \frac{\bar{X}-\mu_0}{\sigma_0/\sqrt{n}}$ with score z. For the alternative

$$H_1 : \begin{cases} \mu < \mu_0, & \text{reject } H_0 \text{ if } z \leq -z_\alpha \\ \mu > \mu_0, & \text{reject } H_0 \text{ if } z \geq z_\alpha \\ \mu \neq \mu_0, & \text{reject } H_0 \text{ if } |z| \geq z_{\alpha/2}. \end{cases} \tag{5.2}$$

Test for μ, σ Unknown, Null $H_0 : \mu = \mu_0$

The SE is $S_X/\sqrt{n}$. The test statistic is $T = \frac{\bar{X}-\mu_0}{S_X/\sqrt{n}}$ with score t. For the alternative

$$H_1 : \begin{cases} \mu < \mu_0, & \text{reject } H_0 \text{ if } t \leq -t(n-1,\alpha) \\ \mu > \mu_0, & \text{reject } H_0 \text{ if } t \geq t(n-1,\alpha) \\ \mu \neq \mu_0, & \text{reject } H_0 \text{ if } |t| \geq t(n-1,\alpha/2). \end{cases} \tag{5.3}$$

Test for σ^2, $\mu = \mu_0$ Known, Null $H_0 : \sigma^2 = \sigma_0^2$

The test statistic is $X^2 = \sum_{i=1}^{n} \left(\frac{X_i-\mu_0}{\sigma_0}\right)^2$ with score χ^2. For the alternative

$$H_1 : \begin{cases} \sigma^2 < \sigma_0^2, & \text{reject } H_0 \text{ if } \chi^2 \leq \chi^2(n,1-\alpha) \\ \sigma^2 > \sigma_0^2, & \text{reject } H_0 \text{ if } \chi^2 \geq \chi^2(n,\alpha) \\ \sigma^2 \neq \sigma_0^2, & \text{reject } H_0 \text{ if } \chi^2 \leq \chi^2(n,1-\alpha/2) \text{ or } \chi^2 \geq \chi^2(n,\alpha/2). \end{cases} \tag{5.4}$$

Test for σ^2, μ Unknown, Null $H_0 : \sigma^2 = \sigma_0^2$

The test statistic is $X^2 = \sum_{i=1}^{n} \left(\frac{X_i-\bar{X}}{\sigma_0}\right)^2 = \frac{n-1}{\sigma_0^2} S_X^2$ with score χ^2. For the alternative

$$H_1 : \begin{cases} \sigma^2 < \sigma_0^2, & \text{reject } H_0 \text{ if } \chi^2 \leq \chi^2(n-1,1-\alpha) \\ \sigma^2 > \sigma_0^2, & \text{reject } H_0 \text{ if } \chi^2 \geq \chi^2(n-1,\alpha) \\ \sigma^2 \neq \sigma_0^2, & \text{reject } H_0 \text{ if } \chi^2 \leq \chi^2(n-1,1-\alpha/2) \text{ or } \chi^2 \geq \chi^2(n-1,\alpha/2). \end{cases} \tag{5.5}$$

Recall that $\chi^2(n,\alpha)$ is the critical value of the χ^2 distribution with n degrees of freedom. The area under the χ^2 pdf to the right of the critical value is α.

We present several examples illustrating these tests.

Example 5.4   Suppose we want to test the hypothesis $H_0 : \mu = 7$ vs. $H_1 : \mu \neq 7$. A random sample of size 25 from a large population yields the sample mean $\bar{x} = 6.3$, with a sample standard deviation of $s_X = 1.2$. Is the difference between the assumed mean and the sample mean real, or is it due to chance? Assume normal populations and a level of significance of $\alpha = 0.05$.

To answer the question, we calculate the value of the test statistic T as

$$t = \frac{\bar{x} - \mu_0}{s_X / \sqrt{n}} = \frac{6.3 - 7}{1.2/5} = -0.116.$$

We are using the T statistic because we do not know the population SD and so must use the sample SD. Since $\alpha = 0.05$, the area below the critical value must be 0.975. The critical value is therefore $t(24, 0.025) = \text{invT}(.975, 24) = 2.064$. For the two-sided test we reject H_0 if $t \notin (-2.064, 2.064)$, but our value of $t = -0.116$ is clearly in the interval. We conclude that we cannot reject the null at the 0.05 level of significance and state that the result in our experiment could be due to chance. Whenever we have such a result, it is said that we **do not reject the null** or that the **null is plausible**. It is never phrased that we accept the alternative. If we want to calculate the p-value we need $P(t(24) \leq -0.116) + P(t(24) \geq 0.116) = 0.9086 > \alpha$, and so there is a large probability we will make an error if we reject the null.

If we had decided to use the one-sided alternative $H_1 : \mu < 7$ on the basis that our observed sample mean is $\bar{x} = 6.3$, our p-value would turn out to be $P(t(24) < -0.116) = 0.4545$, and we still could not reject the null.

Next we present an example in which we have a two-sided test for the variance.

Example 5.5   A certain plumbing supply company produces copper pipe of fixed various lengths and asserts that the standard deviation of the various lengths is 1.3 cm. The mean length is therefore assumed to be unknown. A contractor who is a client of the company decides to test this claim by taking a random sample of 30 pipes and measuring their lengths. The standard deviation of the sample turned out to be $s = 1.1$ cm. The contractor tests the following hypotheses, and sets the level of the test to be $\alpha = 0.01$.

$$H_0 : \sigma^2 = 1.69 \quad \text{vs.} \quad H_1 : \sigma^2 \neq 1.69.$$

Therefore, under an assumption of normality of the lengths, we have the test statistic

$$\chi^2 = \frac{(n-1)s^2}{\sigma_0^2} = \frac{(30-1)(1.21)}{1.69} = 20.763.$$

The χ^2-score is therefore 20.763. According to (5.5), we should reject $H_0 : \sigma^2 = 1.69$ if $\chi^2 \geq \chi^2(29, 0.005) = 52.3$ or $\chi^2 \leq \chi^2(29, 0.995) = 13.1$. Since $\chi^2 = 20.763$, we fail to reject the null hypothesis. There is not enough evidence in the data to reject the company's claim that $\sigma = 1.3$ cm at the $\alpha = 0.01$ level.

Example 5.6   At a certain food processing plant, 36 measurements of the volume of tomato paste placed in cans by filling machines were made. A production manager at the plant thinks that the data supports her belief that the mean is greater than 12 oz. causing lower company profits. Because the sample size $n = 36 > 30$ is large, we assume normality. The sample mean is $\overline{x}_{36} = 12.19$ oz. for the data. The standard deviation is unknown, but the sample standard deviation $s_X = 0.11$. The manager sets up the test of hypotheses.

$$H_0 : \mu = 12 \quad \text{vs.} \quad H_1 : \mu > 12.$$

She sets the level of the test at $\alpha = 0.05$. We have

$$t = \frac{(12.19 - 12)}{0.11/\sqrt{36}} = 10.36.$$

According to (5.3), she should reject $H_0 : \mu = 12$ if $t \geq t(35, 0.05) = 1.69$. The critical region is $[1.69, \infty)$. Since $t = 10.36$, the manager rejects the null. The machines that fill the cans must be recalibrated to dispense the correct amount of tomato paste.

We end this section with a test for a **median**.

Example 5.7   Hypothesis test for the median. We introduce a test for the median m of a continuous random variable X. Recall that the median is often a better measure of centrality of a distribution than the mean. Suppose that $x_1, x_2, \cdots, x_n$ is a random sample of n observed values from X. Consider the test of hypothesis $H_0 : m = m_0$ vs. $H_1 : m > m_0$. Assume that $x_i \neq m_0$ for every i. Let S be the random variable that counts the number of values greater than m_0 (considered successes). If H_0 is true, then by definition of the median, $S \sim \text{Bin}(n, 0.5)$. If S is too large, for example if $S > k$ for some k, H_0 should be rejected. Since the probability of committing a Type I error must be at most α, we have

$$\alpha \geq P(\text{rejecting } H_0 \mid H_0 \text{ is true}) = P(S > k \mid H_0 \text{ is true})$$

$$= 1 - P(S \leq k \mid H_0 \text{ is true}) = 1 - \sum_{i=0}^{k} \binom{n}{i} (0.5)^i (0.5)^{n-i}$$

$$= 1 - \sum_{i=0}^{k} \binom{n}{i} (0.5)^n.$$

Choose the minimum value of k so that $\sum_{i=0}^{k} \binom{n}{i}(0.5)^n \geq 1 - \alpha$.

To illustrate the test, consider the following situation. Non-Hodgkins Lymphoma (NHL) is a cancer that starts in cells called lymphocytes which are part of the body's immune system. Twenty patients were diagnosed with the disease in 2007. The survival time (in weeks) for each of these patients is listed in the table below.

Survival Time (in weeks) for Non-Hogkins lymphoma				
82	665	162	1210	532
476	487	133	129	230
894	449	310	505	55
252	453	630	551	711

We want to test the hypothesis $H_0 : m = 520$ vs. $H_1 : m > 520$ at the $\alpha = 0.05$ level and find the p-value of the test. From the data we observe $S = 7$ data points above 520. Relevant values of Binom$(20, 0.5)$ are shown in the table below.

k	$P(S \leq k \mid H_0$ is true$)$
7	0.1316
...	...
12	0.8684
13	0.9423
14	0.9793
15	0.9941

From this table we see that the smallest k for which $P(X > k \mid H_0$ is true$) \leq \alpha$ is 14. Therefore we reject H_0 if $S > 14$. Since $S = 7$, we retain H_0. As for the p-value we have

$$\text{p-value} = P(S \geq 7) = 1 - P(S < 7) = \sum_{i=0}^{6} \binom{15}{i} \left(\frac{1}{2}\right)^{15} = 0.30362.$$

5.3.2   THE p-VALUE APPROACH TO HYPOTHESIS TESTING

The last section introduced the method of hypothesis testing known as the **critical value approach** which tells us to reject the null when the value of the test statistic is in the critical region. In the introduction we discussed the p-value approach, and we will give further details here. It is important to remember that the two approaches are equivalent, but the p-value approach gives more information to the researcher.

Let $X_1, X_2, \ldots, X_n$ be a random sample from a random variable X with pdf $f_X(x; \vartheta)$. First consider the one-sided test of hypotheses

$$H_0 : \vartheta = \vartheta_0 \quad \text{vs.} \quad H_1 : \vartheta > \vartheta_0.$$

Table 5.1: Conclusions for p-values

p-Value Range	Interpretation
$0 < p \leq 0.01$	Highly significant (very strong evidence against H_0, reject H_0)
$0.01 < p \leq 0.05$	Significant (strong evidence against H_0, reject H_0)
$0.05 < p \leq 0.1$	Weakly significant (weak evidence against H_0, reject H_0)
$0.1 < p \leq 1$	Not significant (insufficient evidence against H_0, do not reject H_0)

Let Θ denote the test statistic under the null hypothesis, that is, when $\vartheta = \vartheta_0$, and let ζ be the value of Θ when the sample data values $X_1 = x_1, X_2 = x_2, \ldots, X_n = x_n$ are substituted for the random variables. The

$$\boxed{\text{p-value of } \zeta \text{ is } P(\Theta \geq \zeta)}$$

which is the probability of obtaining a value of the test statistic Θ as or more extreme than the value ζ that was obtained from the data. For example, suppose that $\Theta \sim Z$ when $\vartheta = \vartheta_0$. We would reject H_0 if $p = P(Z \geq \zeta) < \alpha$, where α is the specified minimum level of significance required to reject the null. We may choose α according to Table 5.1. On the other hand, if $p \geq \alpha$, then we retain, or do not reject, H_0.

The p-value of our observed statistic value ζ determines the **significance** of ζ. If p is the p-value of ζ, then for all test levels α such that $\alpha > p$ we reject H_0. For all test levels $\alpha \leq p$, we retain H_0. Therefore, p can be viewed as the smallest test level for which the null hypothesis is rejected.

If α is not specified ahead of time, we use Table 5.1 to reach a conclusion depending on the resulting p-value.

In a similar way, the alternative $H_1 : \vartheta < \vartheta_0$ is a one-sided test, and the p-value of the observed test statistic ζ is $p = P(\Theta \leq \zeta)$. If $p < \alpha$, we reject the null. The level of significance of the test statistic ζ is the p-value p. Finally, if the alternative hypothesis is two-sided, namely, $H_1 : \vartheta \neq \vartheta_0$, then the p-value is given by

$$\boxed{p = P(\Theta \geq \zeta) + P(\Theta \leq -\zeta).}$$

Remark 5.8   If the distribution of Θ is symmetric, the p-value of a two-sided test is $p = 2P(\Theta \geq \zeta)$, assuming $\zeta \geq 0$. If $\zeta < 0$, its p-value $= 2P(\Theta \leq \zeta)$.

If the distribution of Θ is not symmetric, such as in the χ^2 distribution, we calculate $P(\Theta \geq \zeta)$ as well as $P(\Theta \leq \zeta)$, and the p-value of the two-sided test is $p = 2\min\{P(\Theta \geq \zeta), P(\Theta \leq \zeta)\}$.

Example 5.9   Consider Example 4.9 of the preceding chapter. The hypothesis test is $H_0 : \mu = 98.6°$ vs. $H_1 : \mu \neq 98.6°$. The test statistic is $Z = \frac{\overline{X}-\mu_0}{\sigma_0/\sqrt{n}}$, and the $z-$ score is $z = \frac{98.2-98.6}{0.62/\sqrt{106}} = -6.64$. The two-sided p-value is computed as $P(|Z| \geq 6.64) = 3.14 \times 10^{-11}$ constituting overwhelming evidence against H_0.

Example 5.10   A manufacturer of e-cigarettes (e-cigs) claims that the variance of the nicotine content of its cigarettes is less than 0.77 mg. The sample variance in a random sample of 25 of the company's e-cigs turned out to be 0.41 mg. A health professional would like to know if there is enough evidence to reject the company's claim. The hypothesis test is $H_0 : \sigma^2 = 0.77$ vs. $H_1 : \sigma^2 < 0.77$. Under an assumption of normality, the test statistic is

$$\frac{(n-1)S_X^2}{\sigma_0^2} \sim \chi^2(n-1).$$

Therefore, the χ^2- score is $\chi^2 = \frac{24(0.41)}{0.77} = 12.779$, and so the one-sided p-value is $P(\chi^2(24) \leq 12.779) = 0.0303$. The result is significant evidence against $H_0 : \sigma^2 = 0.77$. We may conclude that there is sufficient evidence to say that the variance is indeed less than 0.77.

If we now change the alternative to $H_1 : \sigma^2 \neq 0.77$, what is the p-value? For that we need to calculate

$$P(\chi^2(24) \leq 12.779) = 0.0303 \quad \text{and} \quad P(\chi^2(24) \geq 12.779) = 0.9696.$$

The p-value is therefore $p = 2\min\{0.0303, 0.9606\} = 0.0606$. Since $p > 0.05$, the result is not statistically significant, and we do not reject the null. This example illustrates how a one-sided alternative may lead to rejection of the null, but a two-sided alternative does not reject the null.

If we use the critical value approach with $\alpha = 0.05$, we calculate the critical values $P(\chi^2(24) \leq a) = 0.025$ which implies $a = 12.401$ and $P(\chi^2(24) \leq b) = 0.975$ which implies $b = 39.364$. The two-sided critical region is $(0, 12.401] \cup [39.364, \infty)$. Since our observed value $\chi^2 = 12.779$ is not in the critical region, we do not reject the null at the 5% level of significance. There is insufficient evidence to conclude that the variance is not 0.77.

5.3.3   TEST OF HYPOTHESES FOR PROPORTIONS

Let X be a Binomial random variable with parameters n and p. Recall that if $np > 5$ and $n(1-p) > 5$, the Central Limit Theorem can be invoked to obtain the approximation

$$\frac{X - np}{\sqrt{np(1-p)}} = \frac{\overline{X} - p}{\sqrt{\frac{p(1-p)}{n}}} = Z \sim N(0, 1).$$

Before deriving the confidence interval for p, the approximate $100(1 - \alpha)\%$ confidence interval

$$\left(\overline{x} - z_{\alpha/2} \sqrt{\frac{p(1 - p)}{n}}, \overline{x} + z_{\alpha/2} \sqrt{\frac{p(1 - p)}{n}} \right)$$

was obtained, but it was noted that this interval cannot be computed from the sample values since p itself appears in the endpoints of the interval. Recall that this problem is solved by replacing p with $\bar{p} = \bar{x}$. However, in a hypothesis test, we have a null $H_0 : p = p_0$ in which we **assume the given population proportion** of p_0, and this is **exactly what we substitute whenever a value of p is required.** Here's how it works.

Consider the two-sided test of hypotheses $H_0 : p = p_0$ vs. $H_1 : p \neq p_0$. We take a random sample and calculate the sample proportion $\overline{p}$. Since $p = p_0$ under the null hypothesis,

$$\frac{\overline{X} - p_0}{\sqrt{\dfrac{p_0(1 - p_0)}{n}}} \sim Z \text{ (approximate)}.$$

This is where we use the assumed value of p under the null. For a given significance level α, we will reject H_0 if

$$\overline{x} = \overline{p} \geq p_0 + z_{\alpha/2} \sqrt{\frac{p_0(1 - p_0)}{n}} \quad \text{or} \quad \overline{x} = \overline{p} \leq p_0 - z_{\alpha/2} \sqrt{\frac{p_0(1 - p_0)}{n}}.$$

The p-value of the test is

$$P \left(|Z| \geq \frac{\overline{p} - p_0}{\sqrt{p_0(1 - p_0)/n}} \right).$$

We reject H_0 if this p-value is less than α.

Example 5.11   One way to cheat at dice games is to use loaded dice. One possibility is to make a 6 come up more than expected by making the 6-side weigh less than the 1-side. In an experiment to test this possible loading, 800 rolls of a die produced 139 occurrences of a 6. Consider the test of hypotheses

$$H_0 : p = \frac{1}{6} \quad \text{vs.} \quad H_1 : p \neq \frac{1}{6}$$

where p is the probability of obtaining a 6 on a roll of the die. Computing the value of the test statistic, we get with $\overline{p} = 139/800 = 0.17375$,

$$z = \frac{\overline{p} - p_0}{\sqrt{\dfrac{p_0(1 - p_0)}{n}}} = \frac{\dfrac{139}{800} - \dfrac{1}{6}}{\sqrt{\dfrac{\dfrac{1}{6}\left(1 - \dfrac{1}{6}\right)}{800}}} = 0.53759.$$

The corresponding p-value is $P(|Z| \geq 0.53759) = 0.5909$. We do not reject H_0, and we do not have enough evidence that the die is loaded. How many 6s need to come up so that we can say it is loaded? Set $\alpha = 0.05$. Then we need $P(|Z| \geq z) < 0.05$ which implies $|z| > 1.96$. Since $z = \dfrac{p - .166}{.01317}$, substitute $p = \frac{n}{800}$, and solve the inequality for n. We obtain that $n \geq 154$ or $n \leq 112$. Therefore, if we roll at least 154 6s or at most 112 6s, we can conclude the die is loaded.

One-sided tests for a proportion are easily constructed. For the tests

$$H_0 : p = p_0 \quad \text{vs.} \quad H_1 : p < p_0 \quad \text{and} \quad H_0 : p = p_0 \quad \text{vs.} \quad H_1 : p > p_0,$$

we reject $H_0 : p = p_0$ if $z \leq -z_\alpha$ and if $z \geq z_\alpha$, respectively. Using the p-value approach, we reject H_0 if the p-value

$$P\left(Z \leq \frac{\bar{p} - p_0}{\sqrt{p_0(1 - p_0)/n}}\right) < \alpha \quad \text{or} \quad P\left(Z \geq \frac{\bar{p} - p_0}{\sqrt{p_0(1 - p_0)/n}}\right) < \alpha.$$

Example 5.12   Suppose an organization claims that 75% of its members have IQs over 135. Test the hypothesis $H_0 : p = 0.75$ vs. $H_1 : p < .75$. Suppose 10 members are chosen at random and their IQs are measured. Exactly 5 had an IQ over 135. Under the null, we have

$$P\left(Z \leq \frac{0.5 - 0.75}{\sqrt{.75 \times .25/10}}\right) = P(Z \leq -1.825) = 0.034.$$

Since $0.03 < 0.05$, our result is statistically significant, and we may reject the null. The evidence supports the alternative.

There is a problem with this analysis in that $np = 10(0.75) = 7.5 > 5$, but $n(1 - p) = 10(0.25) = 2.5 < 5$, so that the use of the normal approximation is debatable. Let's use the exact binomial distribution to analyze this instead. If X is the number of members in a sample of size 10 with IQs above 135, we have $X \sim \text{Binom}(10, 0.75)$ under the null. Therefore, $P(X \leq 5) = \text{binomcdf}(10, 0.75, 5) = 0.078$ is the exact probability of getting 5 or fewer members with IQs below 135. Therefore, we do not have enough evidence to reject the null, and the result is not statistically significant.

5.4   HYPOTHESES TESTS FOR TWO POPULATIONS

In this section we obtain tests of hypotheses for the difference of two **independent samples** $X_1, \ldots, X_m$ and $Y_1, \ldots, Y_n$ from the confidence intervals that were derived in the preceding chapter. **The underlying assumption is that $X_i \sim N(\mu_X, \sigma_X)$ and $Y_j \sim N(\mu_Y, \sigma_Y)$ or else that the samples are large enough that we may approximate by normal populations.** For a given constant d_0 assumed to be the difference in the population means, the tests we consider are of the form

$$H_0 : \mu_X - \mu_Y = d_0 \quad \text{vs. one of} \quad \begin{cases} H_1 : \mu_X - \mu_Y < d_0 \\ H_1 : \mu_X - \mu_Y > d_0 \\ H_1 : \mu_X - \mu_Y \neq d_0. \end{cases}$$

The test statistic depends on the standard deviations of the populations or the samples. Here is a summary of the results. We use the critical value approach to state the results. Computing p-values is straightforward.

SDs of the Populations are Known

The test statistic is $Z = \dfrac{(\bar{X}_m - \bar{Y}_n) - d_0}{\sqrt{\frac{\sigma_X^2}{m} + \frac{\sigma_Y^2}{n}}} \sim N(0, 1)$ when H_0 is true. If z is the score, for the alternative

$$H_1 : \begin{cases} \mu_X - \mu_Y < d_0, & \text{reject } H_0 \text{ if } z \leq -z_\alpha \\ \mu_X - \mu_Y > d_0, & \text{reject } H_0 \text{ if } z \geq z_\alpha \\ \mu_X - \mu_Y \neq d_0, & \text{reject } H_0 \text{ if } |z| \geq z_{\alpha/2}. \end{cases}$$

SDs of the Populations are Unknown, but Equal

The test statistic is $T = \dfrac{(\bar{X}_m - \bar{Y}_n) - d_0}{S_p \sqrt{\frac{1}{m} + \frac{1}{n}}}$ where S_p^2 is the **pooled variance** defined by

$$S_p^2 = \frac{(m - 1)S_X^2 + (n - 1)S_Y^2}{m + n - 2}.$$

When H_0 is true, $T \sim t(m + n - 2)$. If t is the score, for the alternative

$$H_1 : \begin{cases} \mu_X - \mu_Y < d_0, & \text{reject } H_0 \text{ if } t \leq -t(m + n - 2, \alpha) \\ \mu_X - \mu_Y > d_0, & \text{reject } H_0 \text{ if } t \geq t(m + n - 2, \alpha) \\ \mu_X - \mu_Y \neq d_0, & \text{reject } H_0 \text{ if } |t| \geq t(m + n - 2, \alpha/2). \end{cases}$$

SDs of the Populations are Unknown and not Equal

The test statistic is $T = \dfrac{(\bar{X}_m - \bar{Y}_n) - d_0}{\sqrt{\frac{\sigma_X^2}{m} + \frac{\sigma_Y^2}{n}}}$ and is approximately distributed as $t(\nu)$ if H_0 is true where

$$\nu = \left\lfloor \frac{\left(\frac{1}{m}r + \frac{1}{n}\right)^2}{\frac{1}{m^2(m-1)}r^2 + \frac{1}{n^2(n-1)}} \right\rfloor , r = \frac{s_X^2}{s_Y^2}.$$

If t is the score, for the alternative

$$H_1 : \begin{cases} \mu_X - \mu_Y < d_0, & \text{reject } H_0 \text{ if } t \leq -t(v, \alpha) \\ \mu_X - \mu_Y > d_0, & \text{reject } H_0 \text{ if } t \geq t(v, \alpha) \\ \mu_X - \mu_Y \neq d_0, & \text{reject } H_0 \text{ if } |t| \geq t(v, \alpha/2). \end{cases}$$

Paired Samples

If $n = m$ and the two samples $X_1, \ldots, X_n$ and $Y_1, \ldots, Y_n$ are not independent, we consider the difference of the two samples $D_i = X_i - Y_i$ and deem this a one-sample t-test because we assume σ_D is unknown. The test statistic is $T = \frac{\bar{D}_n - d_0}{S_D / \sqrt{n}} \sim t(n-1)$ if H_0 is true. If t is the score, for the alternative

$$H_1 : \begin{cases} \mu_D < d_0, & \text{reject } H_0 \text{ if } t \leq -t(n-1, \alpha) \\ \mu_D > d_0, & \text{reject } H_0 \text{ if } t \geq t(n-1, \alpha) \\ \mu_D \neq d_0, & \text{reject } H_0 \text{ if } |t| \geq t(n-1, \alpha/2). \end{cases}$$

Test for Variances of Two Samples

If we test the relationship between variances of two independent samples, we consider the ratio $r = \frac{\sigma_X^2}{\sigma_Y^2}$. We assume the means μ_X and μ_Y are unknown. We test whether σ_X^2 is a multiple of σ_Y^2 which means we test $H_0 : \frac{\sigma_X^2}{\sigma_Y^2} = r_0$ vs. one of the usual alternatives. The test statistic is $F = \frac{S_X^2}{S_Y^2} \cdot \frac{1}{r_0} \sim F(m-1, n-1)$ when H_0 is true. If f is the score, for the alternative

$$H_1 : \begin{cases} r < r_0, & \text{reject } H_0 \text{ if } f \leq F(m-1, n-1, 1-\alpha) \\ r > r_0, & \text{reject } H_0 \text{ if } f \geq F(m-1, n-1, \alpha) \\ r \neq r_0, & \text{reject } H_0 \text{ if } f \leq F(m-1, n-1, 1-\alpha/2) \text{ or } f \geq F(m-1, n-1, \alpha/2). \end{cases}$$

Example 5.13   In this example, two popular brands of low-fat, Greek-style yogurt, Artemis and Demeter, are compared. Let X denote the weight (in grams) of a container of the Artemis brand yogurt, and let Y be the weight (also in grams) of a container of the Demeter brand yogurt. Assume that $X \sim N(\mu_X, \sigma_X)$ and that $Y \sim N(\mu_Y, \sigma_Y)$. Nine measurements of the Artemis yogurt and thirteen measurements of Demeter yogurt were taken with the following results.

Artemis Brand		
21.7	21.0	21.2
20.7	20.4	21.9
20.2	21.6	20.6

Demeter Brand				
21.5	20.5	20.3	21.6	21.7
21.3	23.0	21.3	18.9	20.0
20.4	20.8	20.3		

Consider the hypotheses

$$H_0 : \frac{\sigma_X^2}{\sigma_Y^2} = 1 \quad \text{vs.} \quad H_1 : \frac{\sigma_X^2}{\sigma_Y^2} \neq 1.$$

We will compute the value of the test statistic

$$F = \frac{S_X^2}{S_Y^2} \cdot \frac{1}{r_0} = \frac{S_X^2}{S_Y^2}$$

since $r_0 = 1$ in this case. From the two samples we compute $s_X^2 = 1.014$ and $s_Y^2 = 0.367$. There-fore, $f = \frac{0.367}{1.014} = 0.3619$.

If we set the level of the test to be $\alpha = 0.05$, then $F(8, 12, 0.025) = 4.20$ and $F(8, 12, 0.975) = 0.28$. Since $0.28 < f < 4.20$, we cannot reject H_0, and we may assume the variances are equal.

Now we want to test if the mean weights of the two brands of yogurt are the same. Con-sider the hypotheses

$$H_0 : \mu_X - \mu_Y = 0 = d_0 \quad \text{vs.} \quad H_1 : \mu_X - \mu_Y \neq 0.$$

The pooled sample standard deviation is computed from the two samples as $s_p = 0.869$. The test statistic is

$$T = \frac{(\overline{X}_m - \overline{Y}_n) - d_0}{S_p \sqrt{\frac{1}{m} + \frac{1}{n}}} = \frac{(\overline{X}_m - \overline{Y}_n)}{S_p \sqrt{\frac{1}{m} + \frac{1}{n}}} \sim t(20).$$

Therefore,

$$t = \frac{(21.03 - 20.89)}{0.869\sqrt{\frac{1}{9} + \frac{1}{13}}} = 0.372.$$

For the same level $\alpha = 0.05$, we reject H_0 if $|t| \geq t(20, 0.025) = 2.086$. Since $t = 0.372$, we retain H_0. Also, we may calculate the p-value as $P(|T| \geq 0.372) = 1 - P(|T| \leq 0.372) = 1 - \text{tcdf}(-0.372, 0.372, 20) = 1 - 0.286 = 0.7138$. We cannot reject the null and conclude there is no difference between the weights of the Artemis and Demeter brand yogurts.

5.4.1 TEST OF HYPOTHESES FOR TWO PROPORTIONS

Recall that we could not construct a confidence interval for the difference $p_X - p_Y$ since we did not know the values of p_X and p_Y. We approximated these values by $\bar{X}_m$ and $\bar{Y}_n$, respectively. Now consider the hypothesis $H_0 : p_X - p_Y = d_0$. If H_0 is true, it follows that

$$\frac{(\bar{X}_m - \bar{Y}_n) - d_0}{\sqrt{\frac{p_X(1-p_X)}{m} + \frac{p_Y(1-p_Y)}{n}}} \sim Z \text{ (approximate)}.$$

If we now approximate p_X and p_Y by $\bar{X}_m$ and $\bar{Y}_n$ respectively, the approximation still holds, and we obtain the test statistic

$$Z = \frac{(\bar{X}_m - \bar{Y}_n) - d_0}{\sqrt{\frac{\bar{X}_m(1-\bar{X}_m)}{m} + \frac{\bar{Y}_n(1-\bar{Y}_n)}{n}}} \sim N(0, 1).$$

So if z is the score, for the alternative

$$H_1 : \begin{cases} p_X - p_Y < d_0, & \text{reject } H_0 \text{ if } z \leq -z_\alpha \\ p_X - p_Y > d_0, & \text{reject } H_0 \text{ if } z \geq z_\alpha \\ p_X - p_Y \neq d_0, & \text{reject } H_0 \text{ if } |z| \geq z_{\alpha/2}. \end{cases}$$

As a special case, suppose that $d_0 = 0$. Therefore, $H_0 : p_X - p_Y = d_0$ becomes $H_0 : p_X = p_Y = p_0$ where p_0 is the common value. Therefore, our test statistic becomes

$$\frac{(\bar{X}_m - \bar{Y}_n)}{\sqrt{\frac{p_0(1-p_0)}{m} + \frac{p_0(1-p_0)}{n}}} = \frac{(\bar{X}_m - \bar{Y}_n)}{\sqrt{p_0(1 - p_0)}\sqrt{\frac{1}{m} + \frac{1}{n}}} \sim Z \text{ (approximate)}.$$

We encounter the same problem as before, namely, that we do not know the value of p_0. Since we are assuming that the population proportions are the same, it makes sense to **pool** the proportions (that is, form a weighted average) as in

$$\bar{P}_0 = \frac{m\bar{X}_m + n\bar{Y}_n}{m + n}.$$

Our test statistic becomes

$$\frac{(\bar{X}_m - \bar{Y}_n)}{\sqrt{\bar{P}_0(1 - \bar{P}_0)}\sqrt{\frac{1}{m} + \frac{1}{n}}} \sim Z \text{ (approximate)}.$$

If z is the score, for the alternative

$$H_1 : \begin{cases} p_X < p_Y, & \text{reject } H_0 \text{ if } z \leq -z_\alpha \\ p_X > p_Y, & \text{reject } H_0 \text{ if } z \geq z_\alpha \\ p_X \neq p_Y, & \text{reject } H_0 \text{ if } |z| \geq z_{\alpha/2}. \end{cases}$$

P-values can be easily computed for all the above tests. As an example, for the test of hypotheses $H_0 : p_X = p_Y$ vs. $H_1 : p_X \neq p_Y$, the p-value is computed as $2P(Z \geq |z|)$ where

$$z = \frac{(\bar{p}_X - \bar{p}_Y)}{\sqrt{\bar{p}_0(1 - \bar{p}_0)}\sqrt{\frac{1}{m} + \frac{1}{n}}}, \quad p_0 = \frac{m\bar{p}_X + n\bar{p}_Y}{m + n}.$$

(Note that $\bar{p}_X = \bar{x}_m$ and $\bar{p}_Y = \bar{y}_n$). The TI calculator command is 2· normalcdf($|z|, \infty$). To compute the absolute value, use the sequence *MATH* → *NUM* → 1.

Example 5.14   A survey was done among boys and girls, ages 7–11, to assess their interest in being part of an effort to colonize the planet Mars as potential settlers. Of 1,900 boys asked this question, 1,311 expressed interest in becoming settlers whereas out of 2,000 girls that were asked the same question, 1,440 said they would be interested. Let p_B and p_G be the proportions of boys and girls, ages 7–11, that are interested in colonizing Mars as settlers. We are interested in determining if these proportions are the same. Consider the test of hypotheses.

$$H_0 : p_B = p_G \quad \text{vs.} \quad H_1 : p_B \neq p_G.$$

Set the level of the test at $\alpha = 0.05$. The pooled estimate of the common proportion p_0 is

$$\bar{p}_0 = \frac{m\bar{p}_B + n\bar{p}_G}{m + n} = \frac{1311 + 1440}{3900} = 0.705.$$

Therefore, the observed value of the test statistic is

$$z = \frac{\dfrac{1311}{1900} - \dfrac{1440}{2000}}{\sqrt{0.705(1 - 0.705)}\sqrt{\dfrac{1}{1900} + \dfrac{1}{2000}}} = -2.053.$$

We reject H_0 since $|z| = 2.053 \geq z_{0.025} = 1.96$. The p-value of $z = -2.053$ is easily calculated to be 0.04.

Example 5.15   We revisit the VIOXX$^{©}$ clinical study introduced at the beginning of the chapter to obtain the solution in terms of the notation and methods of hypothesis testing. Let p_C and p_T denote the true proportions of subjects that experience cardiovascular events (CVs) when taking the drug. Recall that the test of hypotheses was $H_0 : p_T = p_C$ vs. $H_1 : p_T > p_C$. As described in this section, the test statistic is

$$Z = \frac{\bar{T}_{1287} - \bar{C}_{1299}}{\sqrt{\dfrac{\bar{T}_{1287}(1 - \bar{T}_{1287})}{1287} + \dfrac{\bar{C}_{1299}(1 - \bar{C}_{1299})}{1299}}}.$$

For the VIOXX$^{©}$ data, the value is

$$z = \frac{\dfrac{46}{1287} - \dfrac{26}{1299}}{\sqrt{\dfrac{\frac{46}{1287}\left(1 - \frac{46}{1287}\right)}{1287} + \dfrac{\frac{26}{1299}\left(1 - \frac{26}{1299}\right)}{1299}}} = 2.4302.$$

The p-value is calculated as $P(Z \geq 2.4302) = 0.00755$ which is highly significant. It can be interpreted that the probability that the observed difference of means $\bar{T}_{1287} - \bar{C}_{1299}$ is at least

$\frac{46}{1287} - \frac{26}{1299} = 0.0157$ under the assumption that the population means are the same. We reject the null hypothesis. The treatment population proportion appears greater than the control population proportion. What if we pool the proportions since under the null hypothesis, we are assuming that $p_T = p_C$? Does that make a difference? The pooled proportion is $p_0 = \frac{72}{2586}$, and so the value of the test statistic is now

$$z = \frac{\frac{46}{1287} - \frac{26}{1299}}{\sqrt{\frac{72}{2586}\left(1 - \frac{72}{2586}\right)}\sqrt{\frac{1}{1287} + \frac{1}{1299}}} = 2.4305$$

yielding a p-value of $P(Z \geq 2.4305) = 0.00754$. Pooling the proportions has no effect on the significance of the result.

Tables summarizing all the tests for the normal parameters and the binomial proportion are located in Section 5.8.

5.5 POWER OF TESTS OF HYPOTHESES

The **power of a statistical test** refers to its ability to avoid making Type II errors, that is, not rejecting the null hypothesis when it is false. Recall that the probability of making a Type I or Type II error is denoted α and β, respectively. Specifically, we have

$$\alpha = P(\text{rejecting } H_0 \mid H_0 \text{ is true}) \quad \text{and} \quad \beta = P(\text{not rejecting } H_0 \mid H_1 \text{ is true}).$$

In order to quantify the power of a test, we need to be more specific with the alternative hypothesis. Therefore, for each ϑ_1 specified in an alternative $H_1 : \vartheta = \vartheta_1$, we obtain a different value of β. Consequently, β is really a **function** of the ϑ which we specify in the alternative. We define for each $\vartheta_1 \neq \vartheta_0$,

$$\boxed{\beta(\vartheta_1) = P(\text{not rejecting } H_0 \mid H_1 : \vartheta = \vartheta_1 \neq \vartheta_0).}$$

Definition 5.16 The **power of a statistical test at a value** $\vartheta_1 \neq \vartheta_0$ is defined to be $\pi(\vartheta_1) = 1 - \beta(\vartheta_1) = P(\text{rejecting } H_0 \mid H_1 : \vartheta = \vartheta_1 \neq \vartheta_0)$.

The **power of a test is the probability of correctly rejecting a false null**. For reasons that will become clear later, we may define $\pi(\vartheta_0) = \alpha$.

We now consider the problem of how to compute $\pi(\vartheta)$ for different values of $\vartheta \neq \vartheta_0$. To illustrate the procedure in a particular case, suppose we take a random sample of size n from a random variable $X \sim N(\mu, \sigma_0)$. (σ_0 is known.) Consider the set of hypotheses

$$H_0 : \mu = \mu_0 \quad \text{vs.} \quad H_1 : \mu \neq \mu_0.$$

Recall that we reject H_0 if $|\frac{\overline{X} - \mu_0}{\sigma_0/\sqrt{n}}| \geq z_{\alpha/2}$. If, in fact, $\mu = \mu_1 \neq \mu_0$, the probability of a Type II error is

$$\beta(\mu_1) = P(\text{not rejecting } H_0 \mid \mu = \mu_1) \tag{5.6}$$

$$= P\left(\left|\frac{\overline{X} - \mu_0}{\sigma_0/\sqrt{n}}\right| < z_{\alpha/2} \mid \mu = \mu_1\right)$$

$$= P\left(\mu_0 - z_{\alpha/2}\frac{\sigma_0}{\sqrt{n}} < \overline{X} < \mu_0 + z_{\alpha/2}\frac{\sigma_0}{\sqrt{n}} \mid \mu = \mu_1\right)$$

$$= P\left(\frac{(\mu_0 - \mu_1)}{\sigma_0/\sqrt{n}} - z_{\alpha/2} < \frac{\overline{X} - \mu_1}{\sigma_0/\sqrt{n}} < \frac{(\mu_0 - \mu_1)}{\sigma_0/\sqrt{n}} + z_{\alpha/2} \mid \mu = \mu_1\right)$$

$$= P\left(\frac{(\mu_0 - \mu_1)}{\sigma_0/\sqrt{n}} - z_{\alpha/2} < Z < \frac{(\mu_0 - \mu_1)}{\sigma_0/\sqrt{n}} + z_{\alpha/2}\right)\left(\text{since } \frac{\overline{X} - \mu_1}{\sigma_0/\sqrt{n}} \sim Z\right)$$

$$= P\left(\left|Z - \frac{\mu_0 - \mu_1}{\sigma_0/\sqrt{n}}\right| < z_{\alpha/2}\right).$$

We needed to standardize $\overline{X}$ using μ_1 and not μ_0 because μ_1 is assumed to be the correct value of the mean. Once we have $\beta(\mu_1)$, the power of the test at $\mu = \mu_1$ is then $\pi(\mu_1) = 1 - \beta(\mu_1)$.

Remark 5.17   Notice that

$$\lim_{\mu_1 \to \mu_0} \pi(\mu_1) = 1 - P\left(|Z| < z_{\alpha/2}\right) = 1 - (1 - \alpha) = \alpha,$$

and so it makes sense to define $\pi(\mu_0) = \alpha$. Also, note that $\lim_{\mu_1 \to \infty} \pi(\mu_1) = 1$ and $\lim_{\mu_1 \to -\infty} \pi(\mu_1) = 1$.

Keep in mind that the designer of an experiment wants both α and β to be small. We want the power of a test to be close to 1 because the power quantifies the ability of a test to detect a false null.

Example 5.18   A data set of $n = 106$ observations of the body temperature of healthy adults was compiled. The standard deviation is known to be $\sigma_0 = 0.62°$F. (In this example, degrees are Fahrenheit.) Consider the set of hypotheses

$$H_0 : \mu = 98.6° \quad \text{vs.} \quad H_1 : \mu \neq 98.6°.$$

Table 5.2 lists selected values of the power function $\pi(\mu)$ for the above test of hypotheses computed using the formula derived above with the level set at $\alpha = 0.05$. For example, if the alternative is $H_1 : \mu_1 = 98.75°$, the power is $\pi(98.75°) = 0.702$, and so there is about a 70% chance of rejecting a false null if the true mean is 98.75°.

A graph of the power function $\pi(\mu)$ is given below (Fig. 5.1).

Table 5.2: Values of $\pi(\mu_1)$, $\mu_1 \neq 98.6°$

μ_1	98.35	98.40	98.45	98.50	98.55	98.60	98.65	98.70	98.75	98.80	98.85
$\pi(\mu_1)$	0.986	0.913	0.702	0.382	0.132	0.050	0.132	0.382	0.702	0.913	0.986

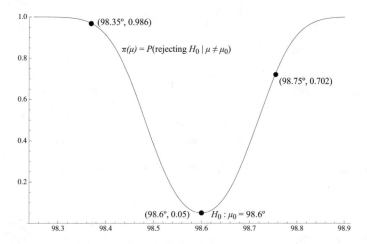

Figure 5.1: A power curve $\pi(\mu)$.

5.5.1   FACTORS AFFECTING POWER OF A TEST OF HYPOTHESES

How can we increase the power of a test of hypotheses? This is an important question since we would like our tests to be as sensitive as possible to the presence of a false null hypothesis. The derivation of the power function $\pi(\mu)$ presented in (5.6) contains important clues. The power function was computed as

$$\pi(\mu_1) = 1 - P\left(\left|Z - \frac{\mu_0 - \mu_1}{\sigma_0/\sqrt{n}}\right| < z_{\alpha/2} \mid \mu_1 \neq \mu_0\right).$$

Assume μ_1, σ_0, and α are all fixed. If the sample size n increases, then $(\mu_0 - \mu_1)/(\sigma_0/\sqrt{n})$ approaches $\pm\infty$ depending upon whether $\mu_0 > \mu_1$ or $\mu_0 < \mu_1$. Consequently,

$$\lim_{n \to \infty} P\left(\left|Z - \frac{\mu_0 - \mu_1}{\sigma_0/\sqrt{n}}\right| < z_{\alpha/2}\right) = 0,$$

and so $\lim_{n \to \infty} \pi(\mu_1) \to 1$. So one way to increase the power of a (two-sided) statistical test is to increase the sample size. This is the most common and practical way of increasing power.

Example 5.19   In the diagram below, power curves are illustrated for two-sided tests for healthy adult body temperature of the form $H_0 : \mu = 98.6°$ vs. $H_1 : \mu \neq 98.6°$, for various choices of sample size.

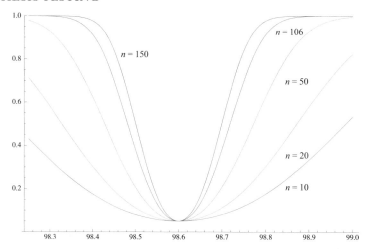

Figure 5.2: Power curves for a two-sided test and different sample sizes.

Suppose we specify the desired power of a test. Can we find the sample size needed to achieve this? Here's an example of the method for a two-sided test if we specify β. We have from (5.6)

$$
\begin{aligned}
1 - \beta &= P\left(\left|Z - \frac{\mu_0 - \mu_1}{\sigma_0/\sqrt{n}}\right| \geq z_{\alpha/2}\right) \\
&= P\left(Z \leq -z_{\alpha/2} - \frac{\mu_0 - \mu_1}{\sigma_0/\sqrt{n}}\right) + P\left(Z \leq -z_{\alpha/2} + \frac{\mu_0 - \mu_1}{\sigma_0/\sqrt{n}}\right) \\
&= P\left(Z \leq -z_{\alpha/2} - \frac{|\mu_0 - \mu_1|}{\sigma_0/\sqrt{n}}\right) + P\left(Z \leq -z_{\alpha/2} + \frac{|\mu_0 - \mu_1|}{\sigma_0/\sqrt{n}}\right).
\end{aligned}
$$

For statistically significant levels α, the first term in the sum above is close to 0, and so it makes sense to equate the power $1 - \beta$ with the second term and solve for n giving us a sample size that is conservative in the sense that the sum of the two terms above will only be slightly larger than $1 - \beta$. Doing this, we get

$$
\begin{aligned}
1 - \beta &= P\left(Z \leq -z_{\alpha/2} + \frac{|\mu_0 - \mu_1|}{\sigma_0/\sqrt{n}}\right) \\
\implies z_\beta &= -z_{\alpha/2} + \frac{|\mu_0 - \mu_1|}{\sigma_0/\sqrt{n}} \\
\implies n &= \left\lceil \left(\frac{\sigma_0(z_\beta + z_{\alpha/2})}{\mu_0 - \mu_1}\right)^2 \right\rceil.
\end{aligned}
$$

If we take $N = \left\lceil \left(\frac{\sigma_0(z_\beta + z_{\alpha/2})}{\mu_0 - \mu_1} \right)^2 \right\rceil$, then for all $n \geq N$, the power at an alternative $\mu_1 \neq \mu_0$ will be at least $1 - \beta$, approaching 1 as $n \to \infty$.

Example 5.20   Continuing with the body temperature example once more, for a test at level $\alpha = 0.05$, suppose we specify that if $|\mu - 98.6°| \geq 0.15°$, then we want $\pi(\mu) \geq 0.90$. By symmetry and monotonicity of the power function, we need only require that $\pi(98.75°) = 0.90$. Therefore, with $z_{\alpha/2} = z_{.025} = 1.96$, $z_\beta = z_{.1} = 1.28$, $\mu_0 = 98.6°$, $\mu_1 = 98.75°$, and $\sigma_0 = 0.62°$, we have a sample size

$$n = \left\lceil \left(\frac{0.62(1.28 + 1.96)}{0.15} \right)^2 \right\rceil = 180.$$

A sample of at least 180 healthy adults must be tested to ensure that the power of the test at all alternatives μ such that $|\mu - 98.6°| \geq 0.15°$ will be at least 0.90.

5.5.2   POWER OF ONE-SIDED TESTS

Computing power and constructing power curves can be done for one-sided tests as well as for two-sided tests. For example, suppose we take a random sample of size n from a random variable $X \sim N(\mu, \sigma_0)$. (σ_0 is known.) Consider the set of hypotheses $H_0 : \mu = \mu_0$ vs. $H_1 : \mu > \mu_0$. At $\mu = \mu_1 \neq \mu_0$,

$$\pi(\mu_1) = 1 - \beta(\mu_1)$$

$$= 1 - P\left(\frac{\overline{X} - \mu_0}{\sigma_0/\sqrt{n}} < z_\alpha \mid \mu = \mu_1 \right)$$

$$= 1 - P\left(\overline{X} < \mu_0 + z_\alpha \frac{\sigma_0}{\sqrt{n}} \mid \mu = \mu_1 \right)$$

$$= 1 - P\left(\frac{\overline{X} - \mu_1}{\sigma_0/\sqrt{n}} < \frac{(\mu_0 - \mu_1)}{\sigma_0/\sqrt{n}} + z_\alpha \mid \mu = \mu_1 \right)$$

$$= 1 - P\left(Z < \frac{(\mu_0 - \mu_1)}{\sigma_0/\sqrt{n}} + z_\alpha \right) \left(\text{since } \frac{\overline{X} - \mu_1}{\sigma_0/\sqrt{n}} \sim Z \right)$$

$$= 1 - P\left(Z < z_\alpha + \frac{\mu_0 - \mu_1}{\sigma_0/\sqrt{n}} \right).$$

We have that $\lim_{\mu_1 \to \mu_0} \pi(\mu_1) = 1 - P(Z < z_\alpha) = 1 - (1 - \alpha) = \alpha$ as before, and so we may define $\pi(\mu_0) = \alpha$. Notice that as $\mu_1 \to -\infty$, $(\mu_0 - \mu_1)/(\sigma/\sqrt{n})$ increases to $+\infty$, and so $\pi(\mu) \to 0$. So for alternative values of $\mu < \mu_0$, the power is less than α, and the smaller μ gets, the less power the test has. This is of no concern really since the test is one-sided, and power only becomes meaningful for alternative values of μ greater than μ_0.

Remark 5.21   As in two-sided tests, power can be increased by increasing sample size. For a one-sided test, the sample size needed for a given β is $n = \left\lceil \left(\frac{\sigma_0(z_\beta+z_\alpha)}{\mu_0-\mu_1} \right)^2 \right\rceil$. (Verify this!)

Example 5.22   The diagram in Figure 5.3 displays power curves for one-sided tests for healthy adult body temperature of the form $H_0 : \mu = 98.6°$ vs. $H_1 : \mu > 98.6°$, for various choices of sample size. Notice that for $\mu < 98.6°$, the power of the test diminishes rapidly.

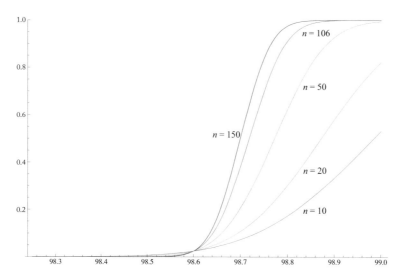

Figure 5.3: One-sided power curves for different sample sizes.

5.6   MORE TESTS OF HYPOTHESES

If we have data from an experiment and we don't know the distribution of the data, how can we test the conjecture that the data comes from, say, a normal distribution, or an exponential distribution? Tests involving the pdf (or pmf in the discrete case) itself and not just the parameters for a known pdf are called **goodness-of-fit** tests. A test of hypotheses of this type has the general form

$$H_0 : f_X(x) = f_{0,X}(x) \quad \text{vs.} \quad H_1 : f_X(x) \neq f_{0,X}(x),$$

where $f_{0,X}(x)$ represents a known pdf.

Another type of test involves testing whether two **traits**, say A and B, are independent or not as in $H_0 : A$ and B are independent vs. $H_1 : A$ and B are dependent. For example, we might want to test whether increasing rates of addiction to opioids (A trait) are related to a downturn in the economy (B trait) or whether the number of hours that a person spends on social media sites (A trait) is related to the person's educational level (B trait).

Both of these classes of tests use essentially the same statistic called Pearson's D statistic introduced by Karl Pearson in 1900. We will describe the distribution of this statistic and show how it is used in hypothesis testing. In preceding sections, we constructed tests for at most two parameters. We will conclude this section with an important multiparameter test that generalizes the two-sample t-test, namely, the **analysis of variance** or **ANOVA** for short.

5.6.1 CHI-SQUARED STATISTIC AND GOODNESS-OF-FIT TESTS

We begin by discussing how the χ^2 distribution is involved in a goodness-of-fit test. Start with the case in which the population is $X \sim \text{Binom}(n, p)$. We know that X is a special case of the multinomial distribution (X_1, X_2) where $X = X_1$ is the number of successes, and X_2 the number of failures in n trials. The probability of a success or a failure is $p_1 = p$ and $p_2 = 1 - p$, respectively. Since X_1 is the sum of n independent Bernoulli trials, we know by the Central Limit Theorem that for large enough n, the following approximations are appropriate:

$$\frac{X_1 - np_1}{\sqrt{np_1(1 - p_1)}} \sim N(0, 1) \quad \text{and} \quad \frac{X_2 - np_2}{\sqrt{np_2(1 - p_2)}} \sim N(0, 1).$$

However, these two distributions are **not independent.** (Why?) We also know from Chapter 2 that $Z \sim N(0, 1)$ implies $Z^2 \sim \chi^2(1)$. Now we calculate

$$
\begin{aligned}
\chi^2(1) \sim D_1 &= \frac{(X_1 - np_1)^2}{np_1(1 - p_1)} \\
&= \frac{(1 - p_1)(X_1 - np_1)^2}{np_1(1 - p_1)} + \frac{p_1(X_1 - np_1)^2}{np_1(1 - p_1)} \\
&= \frac{(X_1 - np_1)^2}{np_1} + \frac{(n - X_2 - np_1)^2}{np_2} \\
&= \frac{(X_1 - np_1)^2}{np_1} + \frac{(n(1 - p_1) - X_2)^2}{np_2} \\
&= \frac{(X_1 - np_1)^2}{np_1} + \frac{(np_2 - X_2)^2}{np_2} \\
&= \frac{(X_1 - np_1)^2}{np_1} + \frac{(X_2 - np_2)^2}{np_2}.
\end{aligned}
$$

Therefore, the sum of the two related distributions is approximately distributed as $\chi^2(1)$. If they were independent, D_1 would be approximately distributed as $\chi^2(2)$. This result can be generalized in a natural way. Specifically, if $(X_1, X_2, \ldots, X_k) \sim \text{Multin}(n, p_1, p_2, \ldots, p_k)$, then it can be shown that as $n \to \infty$,

$$D_{k-1} = \sum_{i=1}^{k} \frac{(X_i - np_i)^2}{np_i} \sim \chi^2(k-1)$$

although the proof is beyond the scope of this text. A common rule of thumb is that n **should be large enough so that $np_i \geq 5$ for each i to guarantee that the approximation is acceptable.** The statistic D_{k-1} is called **Pearson's D statistic.** The subscript $k-1$ is one less than the number of terms in the sum.

We now discuss how the statistic D_{k-1} might be used in testing hypotheses. Consider an experiment with a sample space S, and suppose S is partitioned into disjoint events, i.e., $S = \cup_{i=1}^{k} A_i$ where $A_1, A_2, \ldots, A_k$ are mutually disjoint subsets such that $P(A_i) = p_i$. Clearly, $p_1 + p_2 + \cdots + p_k = 1$. If the experiment is repeated independently n times, and X_i is defined to be the number of times that A_i occurs, then $(X_1, X_2, \ldots, X_k) \sim \text{Multin}(n, p_1, p_2, \ldots, p_k)$. We want to know if the experimental results for the proportion of time event A_i occurs matches some prescribed proportion $p_{0,i}$ for each $i = 1, 2, \ldots, k$. The test of hypotheses for given probabilities $p_{0,1}, \ldots, p_{0,k}$ is

$$H_0 : p_1 = p_{0,1}, \, p_2 = p_{0,2}, \, \ldots, \, p_k = p_{0,k} \quad \text{vs.}$$
$$H_1 : p_i \neq p_{0,i} \text{ for at least one } i.$$

This is now setup for the χ^2 goodness-of-fit test. If H_0 is true, then $(X_1, X_2, \ldots, X_k) \sim \text{Multin}(n, p_{0,1}, p_{0,2}, \ldots, p_{0,k})$, and for large enough n, D_{k-1} is distributed approximately as

$$D_{k-1} = \sum_{i=1}^{k} \frac{(X_i - np_{0,i})^2}{np_{0,i}} \sim \chi^2(k-1).$$

Since each X_i represents the **observed** frequency of the observations in A_i and $np_{0,i}$ is the **expected** frequency of the observations in A_i, if the null hypothesis is true, we would expect the value of D_{k-1} to be small. We should **reject the null hypothesis if the value of D_{k-1} appears to be too large.** To make sure that we commit a Type I error at most $100\alpha\%$ of the time, we need

$$P(D_{k-1} \geq \chi^2(k-1, \alpha)) = P\left(\sum_{i=1}^{k} \frac{(X_i - np_{0,i})^2}{np_{0,i}} \geq \chi^2(k-1, \alpha)\right) = \alpha.$$

The value of the test statistic D_{k-1} with observations $X_1 = x_1, \ldots, X_n = x_n$, is

$$d_{k-1} = \sum_{i=1}^{k} \frac{(x_i - np_{0,i})^2}{np_{0,i}}.$$

Example 5.23   A tetrahedral die is tossed 68 times to determine if the die is fair or not. The sides are labeled 1–4. Consider the following set of hypotheses.

$$H_0 : \text{the die is fair} \quad \text{vs.} \quad H_1 : \text{the die is not fair.}$$

The observed and expected frequencies are listed in the table below.

Side	Observed Frequency	Expected Frequency
1	22	$0.25 \cdot 68 = 17.0$
2	15	17.0
3	19	17.0
4	12	17.0

The value of D_3 is

$$d_3 = \frac{(22 - 17)^2}{17} + \frac{(15 - 17)^2}{17} + \frac{(19 - 17)^2}{17} + \frac{(12 - 17)^2}{17} = 3.4118.$$

Set $\alpha = 0.05$. Since $d_3 < \chi^2(3, 0.05) = 7.815$, the null hypothesis is not rejected, and there is not enough evidence to claim that the die is not fair. In addition, the p-value of the test is $P(D_3 \geq 3.4118) = 0.3324 > 0.05$.

Example 5.24   The printer's proofs of a new 260-page probability and statistics book contains typographical errors (or not) on each page. The number of pages on which i errors occurred is given in the table below.

Num of Errors	0	1	2	3	4	5	≥ 6
Num of Pages	77	90	55	30	5	3	0

Suppose we conjecture that these 260 values resulted from sampling a Poisson random variable X with $\lambda = 2$. The random variable $X \sim \text{Poisson}(\lambda)$ if $P(X = i) = e^{-\lambda}\frac{\lambda^i}{i!}$, $i = 0, 1, 2, \ldots$. The hypothesis test is

$$H_0 : \text{data is Poisson with } \lambda = 2 \text{ vs. } H_1 : \text{data is not Poisson with } \lambda = 2.$$

To apply our method we must first compute the probabilities $P(X = i)$, $i = 0, \ldots, 5$, and also $P(X \geq 6)$ since the Poisson random variable takes on every nonnegative integer value. The results are listed below.

i	Probability $e^{-2}\frac{2^i}{i!}$	Expected Frequency $260 \cdot P(X = i)$
0	0.13534	35.188
1	0.27067	70.374
2	0.27067	70.374
3	0.18045	46.917
4	0.09022	23.457
5	0.03609	9.383
≥ 6	$1 - P(X \leq 5) = 0.01656$	4.306

Since the expected frequency of at least 6 errors is less than 5, we must combine the entries for $i = 5$ and $i \geq 6$. Our revised table is displayed below.

i	Probability	Expected Frequency
0	0.13534	35.188
1	0.27067	70.374
2	0.27067	70.374
3	0.18045	46.917
4	0.09022	23.457
≥ 5	0.05265	13.689

The value of D_5 is

$$d_5 = \frac{(77 - 35.188)^2}{35.188} + \frac{(90 - 70.374)^2}{70.374} + \frac{(55 - 70.374)^2}{70.374} +$$

$$\frac{(30 - 46.917)^2}{46.917} + \frac{(5 - 23.457)^2}{23.457} + \frac{(3 - 13.689)^2}{13.689} = 87.484.$$

Let $\alpha = 0.01$. Since $d_5 \geq \chi^2(5, 0.01) = 15.086$, we reject H_0. Another way to see this is by calculating the p-value $P(D_5 \geq 87.484) = 2.26 \times 10^{-17} < 0.01$. It is extremely unlikely that the data is generated by a Poisson random variable with $\lambda = 2$.

Example 5.25   Continuing with the previous example, you may question why we took $\lambda = 2$. Actually it was just a guess. A better way is to estimate the value of λ directly from the data instead of attempting to guess it. Since the expectation of a Poisson random variable X with parameter λ is $E(X) = \lambda$, we can estimate λ from the sample mean as

$$\bar{\lambda} = \bar{x}_{260} = \frac{77 \cdot 0 + 90 \cdot 1 + 55 \cdot 2 + 30 \cdot 3 + 5 \cdot 4 + 3 \cdot 5}{260} = 1.25.$$

The estimated expected frequencies with this value of λ are listed below.

i	$P(X = i) = e^{-1.25}\frac{1.25^i}{i!}$	Expected Frequency $260 \cdot P(X = i)$
0	0.2865	74.490
1	0.3581	93.106
2	0.2238	58.188
3	0.0933	24.258
4	0.0291	7.566
5	0.0073	1.898
≥ 6	$1 - P(X \leq 5) = 0.0019$	0.494

We must combine the entries for $i = 4$, $i = 5$, and $i \geq 6$ to replace the bottom three rows with $i \geq 4$, $P(X \geq 4) = 0.0383$, and expected frequency 9.958.

Since one of the parameters, λ, had to be estimated using the sample values, it turns out that under the null hypothesis, the D statistic loses a degree of freedom. That is, $D_4 \sim \chi^2(3)$. The value of D_4 is now computed as

$$d_4 = \frac{(77 - 74.490)^2}{74.490} + \frac{(90 - 93.106)^2}{93.106} + \frac{(55 - 58.188)^2}{58.188}$$
$$+ \frac{(30 - 24.258)^2}{24.258} + \frac{(8 - 9.958)^2}{9.958} = 2.107.$$

Take $\alpha = 0.01$. Since $d_4 < \chi^2(3, 0.01) = 11.345$ or, calculating $P(D_4 \geq 2.107) = 0.55049 > \alpha$, we do not reject H_0. It is plausible that the population is described by a Poisson random variable with $\lambda = 1.25$.

Remark 5.26   The previous example addresses an important issue. **If any of the parameters of the proposed distribution must be estimated, then the D statistic under H_0 loses degrees of freedom.** In particular, if the proposed distribution has r parameters that must be estimated from the data, then $D_{k-1} \sim \chi^2(k - 1 - r)$ under the null hypothesis. The proof is fairly involved and therefore is omitted.

Example 5.27   Simulation on a computer of a random variable depends on generating random numbers in $[0, 1]$. Since computers are deterministic and not stochastic, any algorithm generating pseudorandom numbers must be tested to see if it actually produces pseudorandom samples from the uniform random variable on the interval $[0, 1]$. On one run of such a program, the following set of 50 numbers was generated.

Pseudorandom Numbers from [0, 1]				
0.00418818	0.489868	0.860478	0.531297	0.134487
0.299301	0.0126372	0.00376535	0.0380281	0.00198181
0.640423	0.54803	0.51956	0.143379	0.0504356
0.636537	0.136595	0.229917	0.211021	0.0756147
0.791362	0.651962	0.726103	0.986798	0.128636
0.474316	0.491401	0.693047	0.188199	0.14045
0.47789	0.38617	0.00837938	0.198406	0.33602
0.229248	0.46666	0.398312	0.340956	0.00351918
0.168613	0.858063	0.240602	0.347544	0.155587
0.0405615	0.427129	0.963142	0.886288	0.0283893

How do we know these numbers are really random, i.e., generated correctly from $X \sim$ Unif[0, 1]? Here's how to apply a goodness-of-fit test for continuous distributions. First, partition the range of the random variable in some way, for example, into equal-length subintervals. Let $A_1 = [0, 0.1]$, and for $2 \le i \le 10$, let $A_i = (0.1(i-1), 0.1i]$. The frequencies of the numbers in each of the intervals A_i are given in the following table. The expected frequencies are calculated assuming they do come from Unif[0, 1].

Interval	Frequency	Expected Frequency
[0.0, 0.1]	11	5
(0.1, 0.2]	9	5
(0.2, 0.3]	5	5
(0.3, 0.4]	5	5
(0.4, 0.5]	6	5
(0.5, 0.6]	3	5
(0.6, 0.7]	4	5
(0.7, 0.8]	2	5
(0.8, 0.9]	3	5
(0.9, 1.0]	2	5

We consider the test of hypotheses

$$H_0 : \text{data generated from Unif}[0, 1] \quad \text{vs.} \quad H_1 : \text{data not generated from Unif}[0, 1].$$

The value of $D_9 \sim \chi^2(9)$ is computed as

$$d_9 = \frac{(11-5)^2}{5} + \frac{(9-5)^2}{5} + \frac{(5-5)^2}{5} + \frac{(5-5)^2}{5} + \frac{(6-5)^2}{5}$$
$$+ \frac{(3-5)^2}{5} + \frac{(4-5)^2}{5} + \frac{(2-5)^2}{5} + \frac{(2-5)^2}{5} + \frac{(2-5)^2}{5} = 17.0.$$

Let $\alpha = 0.05$. Since $P(D_9 \geq 17) = 0.0487 < \alpha$ or $\chi^2(9, 0.05) = 16.919 < 17$, we reject H_0 that the data is from the uniform distribution on $[0, 1]$. Notice that whether to reject the null hypothesis is unclear, and perhaps the experiment should be repeated. The results are statistically significant but not highly significant.

5.6.2   CONTINGENCY TABLES AND TESTS FOR INDEPENDENCE

Consider an experiment in which outcomes have *two* traits, say A and B. For example, A might denote yearly income and B educational level. Within each trait, there are a finite number of **mutually exclusive** categories. Yearly income might be categorized as low, middle, or high, and educational level as high school, some college, bachelor's degree, some post-graduate education, or graduate degree. How do we determine if trait A is independent of trait B? To generalize, let $A_1, A_2, \ldots, A_r$ denote the mutually exclusive categories within trait A, and let $B_1, B_2, \ldots, B_c$ denote the categories within B. The trait A can be viewed as a random variable with r categorical responses (the A_i), and likewise for B with c responses. In this view, we are asking if the random variables A and B are independent. Suppose that in addition to letting A denote the trait, we also let A denote the *event* of the experiment containing all outcomes with trait A, and similarly for B. Let

$$P(A_i \cap B_j) = p_{ij}.$$

Perform the experiment n times, and let X_{ij} be the frequency of the occurrence (observation) of the event $A_i \cap B_j$. The values of X_{ij} are usually displayed in what is referred to as a **contingency table**. Specifically, X_{ij} is placed in row i and column j. The general form of a contingency table is displayed below.

	Categories	B_1	B_2	$\cdots$	B_{c-1}	B_c	Row Totals
				Trait B			
Trait A	A_1	X_{11}	X_{12}	$\cdots$	$X_{1(c-1)}$	X_{1c}	R_1
	A_2			$\cdots$			R_2
	$\vdots$	$\vdots$	$\vdots$	$\ddots$	$\vdots$	$\vdots$	$\vdots$
	A_{r-1}			$\cdots$			R_{r-1}
	A_r	X_{r1}	X_{r2}	$\cdots$	$X_{r(c-1)}$	X_{rc}	R_r
	Column Totals	C_1	C_2	$\cdots$	C_{c-1}	C_c	n

The adjective "contingent" is often used to describe the situation when an event can occur only when some other event occurs first. For example, earning a high salary in the current economy is contingent upon finding a high-tech job. In this sense, a contingency table's function is to reveal some type of dependency or relatedness between the traits.

We know that

$$(X_{11}, X_{12}, \ldots, X_{1c}, \ldots, X_{r1}, \ldots, X_{rc}) \sim \text{Multin}(n, p_{11}, p_{12}, \ldots, p_{1c}, \ldots, p_{r1}, \ldots, p_{rc}),$$

and for large enough n, D_{rc-1} is approximately distributed as

$$D_{rc-1} = \sum_{i=1}^{r} \sum_{j=1}^{c} \frac{(X_{ij} - np_{ij})^2}{np_{ij}} \sim \chi^2(rc - 1).$$

This result can be used in a test of hypotheses for the **independence** of the traits (random variables) A and B. Clearly, traits A and B are independent if $P(A_i \mid B_j) = P(A_i)$ for every i and j. We know that independence of events, in this case traits, is equivalent to

$$P(A_i \cap B_j) = P(A_i|B_j)P(B_j) = P(A_i)P(B_j)$$

for every i and j. Consider now the hypotheses

$$H_0 : P(A_i \cap B_j) = P(A_i)P(B_j), i = 1, \ldots, r, j = 1, \ldots, c \text{ vs.}$$
$$H_1 : P(A_i \cap B_j) \neq P(A_i)P(B_j) \text{ for some } i \text{ and } j.$$

If traits A and B are independent, the null should hold. Again, the null is formulated to assume independence because otherwise we have no way to account for any level of dependence.

Let the row frequencies and column frequencies be denoted, respectively, by $p_{i*} = P(A_i)$ and $p_{*j} = P(B_j)$. The set of hypotheses above can now be rewritten more compactly as

$$H_0 : p_{ij} = p_{i*}p_{*j}, i = 1, \ldots, r, j = 1, \ldots, c \text{ vs.}$$
$$H_1 : p_{ij} \neq p_{i*}p_{*j} \text{ for some } i \text{ and } j.$$

In other words, the null states that the probability of each cell should be the product of the corresponding row and column probabilities. Under the null hypothesis, D_{rc-1} is approximately distributed as

$$D_{rc-1} = \sum_{i=1}^{r} \sum_{j=1}^{c} \frac{(X_{ij} - np_{i*}p_{*j})^2}{np_{i*}p_{*j}} \sim \chi^2(rc - 1).$$

The problem is that we do not know any of the probabilities involved here. Our only course of action is to estimate them from the sample values. Defining

$$X_{i*} = \sum_{j=1}^{c} X_{ij} = \text{row sum} \quad \text{and} \quad X_{*j} = \sum_{i=1}^{r} X_{ij} = \text{column sum},$$

we can estimate p_{i*} and p_{*j} as

$$\hat{p}_{i*} = \frac{X_{i*}}{n} = \frac{\text{row sum}}{n} \quad \text{and} \quad \hat{p}_{*j} = \frac{X_{*j}}{n} = \frac{\text{column sum}}{n}$$

where n is the total number of observations. Since we have estimated unknown parameters, the number of degrees of freedom of the D statistic is reduced. Once we estimate $\hat{p}_{ij}, j = 1, \ldots, c -$

1, then $\hat{p}_{ic}$ is fixed. Similarly, $\hat{p}_{rj}$ is fixed. So exactly $(r-1)+(c-1)=r+c-2$ parameters must be estimated, reducing the number of degrees of freedom by $r+c-2$. But $rc-1-(r+c-2)=rc-r-c+1=(r-1)(c-1)$. Therefore, D_{rc-1} is approximately distributed as

$$D_{rc-1}=\sum_{i=1}^{r}\sum_{j=1}^{c}\frac{(X_{ij}-n\hat{p}_{i*}\hat{p}_{*j})^2}{n\hat{p}_{i*}\hat{p}_{*j}} \sim \chi^2((r-1)(c-1)).$$

If traits A and B are really independent, we would expect the value of D_{rc-1} to be small. As before, we will reject H_0 at level α if the value of D_{rc-1} is too large, specifically, if the value is at least $\chi^2((r-1)(c-1),\alpha)$. If we use the p-value approach, we would compute

$$P(D_{rc-1}\geq d_{rc-1})=\chi^2\text{cdf}(d_{rc-1},\infty,(r-1)(c-1)).$$

Finally, as a rule of thumb, the estimated expected frequencies should be at least 5 for each i and j. If not, rows and columns should be collapsed to achieve this requirement.

Example 5.28   Data is collected to determine if political party affiliation is related to whether or not a person opposes, supports, or is indifferent to water use restrictions in a certain Southwestern American city that is experiencing a severe drought. A total of 500 adults who belonged to one of the two major political parties were contacted in the survey. We would like to know if a person's party affiliation and his or her opinion about water restrictions are related. The hypotheses to be tested are

H_0 : party affiliation and water use restriction opinion are independent vs.
H_1 : party affiliation and water use restriction opinion are dependent.

The results of the survey are presented in the following contingency table.

		Response			
	Categories	Approves (A)	Opposes (O)	Indifferent (I)	Row Totals
Party	Democrat (D)	138 115.14	64 84.36	83 85.5	285
Affiliation	Republican (R)	64 86.86	84 63.64	67 64.5	215
	Column Totals	202	148	150	500

The estimated probabilities are

$$\hat{p}_{D*}=\frac{285}{500}=0.57, \quad \hat{p}_{R*}=\frac{215}{500}=0.43, \quad \hat{p}_{*A}=\frac{202}{500}=0.404,$$

$$\hat{p}_{*O}=\frac{148}{500}=0.296, \quad \hat{p}_{*I}=\frac{150}{500}=0.3.$$

The estimated expected frequencies in the table (displayed below the number of observations) are calculated by taking $500\times$ row prob. $\times$ col. prob. For example, the expected frequency of

Democrats who approve is $500\hat{p}_{D*}\hat{p}_{*A} = 500 \cdot 0.57 \cdot 0.404 = 115.14$. The value of $D_{2 \cdot 3 - 1} = D_5$ is calculated as

$$d_5 = \sum_{i \in \{D,R\}} \sum_{j \in \{A,O,I\}} \frac{(X_{ij} - n\hat{p}_{i*}\hat{p}_{*j})^2}{n\hat{p}_{i*}\hat{p}_{*j}}$$

$$= \frac{(138 - 115.14)^2}{115.14} + \frac{(64 - 84.36)^2}{84.36} + \frac{(83 - 85.5)^2}{85.5}$$

$$+ \frac{(64 - 86.86)^2}{86.86} + \frac{(84 - 63.64)^2}{63.64} + \frac{(67 - 64.5)^2}{64.5} = 22.135.$$

Because of the estimated parameters, $D_5 \sim \chi^2(2)$. Set $\alpha = 0.05$. The p-value of 22.135 is $P(D_5 \geq 22.135) = 0.000016 < \alpha$ constituting strong evidence against H_0. A person's opinion on water use restrictions during the current drought appears to be dependent on political party affiliation.

Remark 5.29   A note of caution is in order here. The statistic D_{rc-1} is discrete, and we are using a continuous distribution, namely the χ^2 distribution, to approximate it. If D_{rc-1} is approximated by $\chi^2(1)$ (for example, for a 2×2 contingency table) or when at least one of the estimated expected frequencies is less than 5, a **continuity correction** has been suggested to improve the approximation just as we do in using a normal distribution to approximate a binomial distribution. The suggestion is that the D statistic be corrected as

$$D_{rc-1} = \sum_{i=1}^{r} \sum_{j=1}^{c} \frac{(|X_{ij} - n\hat{p}_{i*}\hat{p}_{*j}| - 0.5)^2}{n\hat{p}_{i*}\hat{p}_{*j}}.$$

This correction has a tendency however to over-correct and may lead to larger Type II errors.

Example 5.30   A sample of 185 prisoners who experienced trials in a certain criminal jurisdiction was taken, and the results are presented in the 2×2 contingency table below.

		Verdict		
	Categories	Acquitted (A)	Convicted (C)	Row Totals
Offender	Female (F)	39 39.495	5 4.505	44
Gender	Male (M)	127 126.45	14 14.55	141
	Column Totals	166	19	185

The estimated probabilities are

$$\hat{p}_{F*} = \frac{44}{185} = 0.238, \hat{p}_{M*} = \frac{141}{185} = 0.762,$$

$$\hat{p}_{*A} = \frac{166}{185} = 0.897, \hat{p}_{*C} = \frac{19}{185} = 0.103.$$

Since one of the estimated expected frequencies is less than 5, and we are working with a 2×2 contingency table, we will use the continuity correction. We compute

$$d_3 = \frac{(|39 - 39.495| - 0.5)^2}{39.495} + \frac{(|5 - 4.505| - 0.5)^2}{4.505}$$

$$+ \frac{(|127 - 126.450| - 0.5)^2}{126.450} + \frac{(|14 - 14.550| - 0.5)^2}{14.550} = 0.000198.$$

If $\alpha = 0.05$, then $\chi^2(1, 0.05) = 3.841$. We fail to reject H_0. The gender of the offender and whether or not the offender is convicted or acquitted appear not to be related.

Test for Homogeneity

We end our treatment of χ^2- tests with a **test for homogeneity**. In our discussion of the test for independence, we let $A_1, A_2, \ldots, A_r$ denote the mutually exclusive categories within trait A, and $B_1, B_2, \ldots, B_c$ denote the categories within B. Recall that the trait A can be viewed as a random variable with r categorical responses (the A_i), and likewise for B with c responses. A single random sample of size n consists of n pairs of these categorical responses, the first one coming from A and the second from B. The data is assembled in a $r \times c$ contingency table. In a test for homogeneity, what we are interested in is whether for each category A_i of the trait A, the distribution of the B responses is the same. The following example should clarify the question of interest.

Example 5.31   The rash of school shootings across the USA has prompted the call for armed guards in schools. A poll was conducted across five Midwestern states: Indiana, Illinois, Michigan, Wisconsin, and Ohio. It was decided that a random sample of 100 parents in each state should be asked whether they approved or disapproved of armed guards in the schools. The results of the poll are given below in a 5×2 contingency table.

	Categories	Approves (A)	Dissapproves (D)	Row Totals
		Approval/Dissapproval		
	IN	65	35	100
	IL	71	29	100
State	MI	78	22	100
	WI	82	18	100
	OH	70	30	100
	Column Totals	366	134	500

The A category is the state, the B category the approval/disapproval. We are interested in asking whether the percentage of parents in each of the Midwestern states that approve or dissaprove of armed gaurds in the schools is the same. If so, we would say that the distribution of approval/dissapproval across the states is the same.

We now develop the test of homogeneity. Before starting, a remark is in order. Typically, the test of homogeneity is developed by taking random samples from distinct populations (the A categories) whose sizes are determined **a priori**. (This is the case in the example above.) That is, the sizes of these random samples are determined before sampling is done. In our presentation, a single sample of size n is taken from a single population. The effect of doing this is that the sizes of these random samples restricted to each category **are themselves random**. However, it can be proved that the test of homogeneity derived under the assumption that the samples are taken from distinct populations results in exactly the same test that is described here. The technical details are not of immediate interest and are omitted.

As in our discussion of testing independence, suppose that in addition to letting A_i denote the category, we also let A_i denote the *event* in the experiment containing all outcomes within category A_i, and similarly for B_j. As before, let $p_{ij} = P(A_i \cap B_j)$, $p_{i*} = P(A_i)$, and $p_{*j} = P(B_j)$. The null hypothesis states that the distribution of the B categories be the same across all the A categories. Formally,

$$H_0 : \text{for each } i, P(B_j \mid A_i) = P(B_j) = p_{*j} \text{ for every } j \quad \text{vs.}$$
$$H_1 : \text{some } i, P(B_j \mid A_i) \neq P(B_j) = p_{*j} \text{ for some } j.$$

Therefore, under H_0,

$$p_{ij} = P(A_i \cap B_j) = P(B_j \mid A_i)P(A_i) = P(B_j)P(A_i) = p_{i*}p_{*j}.$$

These probabilities must be approximated by the data in the contingency table. If X_{i*} is the observed frequence of category A_i (row sum), X_{*j} is the observed frequency of category B_j (column sum), and n is the total number of observations, then

$$\hat{p}_{i*} = \frac{X_{i*}}{n} = \frac{\text{row sum}}{n} \quad \text{and} \quad \hat{p}_{*j} = \frac{X_{*j}}{n} = \frac{\text{column sum}}{n}.$$

Therefore, under H_0, if X_{ij} is the observed frequence of $A_i \cap B_j$,

$$\hat{p}_{ij} = n\hat{p}_{i*}\hat{p}_{*j} = \frac{\text{row sum} \times \text{column sum}}{n}.$$

The above analysis results in the same test statistic as that derived for the test of independence, namely,

$$D_{rc-1} = \sum_{i=1}^{r}\sum_{j=1}^{c} \frac{(X_{ij} - n\hat{p}_{i*}\hat{p}_{*j})^2}{n\hat{p}_{i*}\hat{p}_{*j}} \sim \chi^2\left((r-1)(c-1)\right) \text{ (approximate)}.$$

We now return to the example introduced above.

Example 5.32 (Example 5.31 Continued) The estimated probabilities are

$$\hat{p}_{*A} = \tfrac{366}{500} = 0.732, \hat{p}_{*D} = \tfrac{134}{500} = 0.268,$$
$$\hat{p}_{IN*} = \hat{p}_{IL*} = \hat{p}_{MI*} = \hat{p}_{WI*} = \hat{p}_{OH*} = 0.2.$$

The estimated expected frequencies are listed in the updated contingency table below.

	Categories	Approval/Dissapproval		Row Totals
		Approves (A)	Dissapprove (D)	
State	IN	65 73.2	35 26.8	100
	IL	71 73.2	29 26.8	100
	MI	78 73.2	22 26.8	100
	WI	82 73.2	18 26.8	100
	OH	70 73.2	30 26.8	100
	Column Totals	366	134	500

The value of $D_{2.5-1} = D_9$ is approximately distributed as $\chi^2(4)$. Computing the value of D_9 we get

$$d_9 = \sum_{i \in \{A,D\}} \sum_{j \in \{IN,IL,MI,WI,OH\}} \frac{(X_{ij} - n\hat{p}_{i*}\hat{p}_{*j})^2}{n\hat{p}_{i*}\hat{p}_{*j}}$$

$$= \frac{(65 - 73.2)^2}{73.2} + \frac{(35 - 26.8)^2}{26.8} + \frac{(71 - 73.2)^2}{73.2} + \frac{(29 - 26.8)^2}{26.8} +$$
$$\frac{(78 - 73.2)^2}{73.2} + \frac{(22 - 26.8)^2}{26.8} + \frac{(82 - 73.2)^2}{73.2} + \frac{(18 - 26.8)^2}{26.8} +$$
$$\frac{(70 - 73.2)^2}{73.2} + \frac{(30 - 26.8)^2}{26.8}$$
$$= 9.3182.$$

Set $\alpha = 0.01$. The p-value is $P(D_9 \geq 9.3182) = 0.05362 > \alpha$. We retain the null. The distribution of parents across the five Midwestern states that approve and dissaprove having armed guards in the schools appear to be the same.

Remark 5.33   The TI calculator can perform the χ^2- test. Enter the observed values in a list, say L_1, and the expected values in a second list L_2. Press *STAT* $\to$ *TESTS* $\to$ χ^2GOF – *Test*. The calculator will return the value of the χ^2 statistic as well as the p-value. In addition, it will return the vector of each term's contribution to the statistic so that it can be determined which terms contribute the most to the total. For a test of independence/homogeneity, the observed values and expected values can be entered into matrices A and B. The calculator's χ^2GOF test will return the value of the statistic as well as the p-value.

5.6.3   ANALYSIS OF VARIANCE

Earlier in the chapter, we presented a method to test the set of hypotheses $H_0 : \mu_X = \mu_Y$ vs. $H_1 : \mu_X \neq \mu_Y$ based on the Student's t random variable. In this section, we will describe a procedure that generalizes the two-sample test to $k \geq 2$ samples. Specifically, there are j treatments resulting in outcomes $X_j \sim N(\mu_j, \sigma)$, and suppose that $X_{1j}, X_{2j}, \ldots, X_{n_j j}$ is a random sample of size n_j from X_j, $j = 1, 2, \ldots, k$. We will assume that the random samples are independent of one another, and the **variances are the same for all the random variables** X_j, $j = 1, 2, \ldots, k$, with common value σ^2. The hypothesis test that determines if the treatments result in the same means or if there is at least one difference is

$$H_0 : \mu_1 = \mu_2 = \cdots = \mu_k \quad \text{vs.}$$
$$H_1 : \mu_j \neq \mu_{j'} \text{ for some } j \neq j'.$$

The test for this set of hypotheses will be based on the F random variable as opposed to the Student's t when there are only two treatments. The tests will be equivalent in the case $k = 2$. The development which follows has traditionally been referred to as the **analysis of variance** (or **ANOVA** for short), and is part of the statistical theory of the design of experiments. In the next chapter we will also present an ANOVA in connection with linear regression.

We now introduce some specialized notation that is traditionally used in ANOVA. Let $n = n_1 + n_2 + \cdots + n_k$, the total number of random variables across the k random samples. In addition, let

$$\mu = \frac{1}{n} \sum_{j=1}^{k} n_j \mu_j = \sum_{j=1}^{k} \frac{n_j}{n} \mu_j$$

which is clearly a weighted average of all the means μ_j, $j = 1, 2, \ldots, k$, since $\sum_{j=1}^{k} \frac{n_j}{n} = 1$. If we assume the null hypothesis, then $\mu = \mu_1 = \cdots = \mu_k$, and μ is the common mean. Finally, let

$$\overline{X}_{*j} = \frac{1}{n_j} \sum_{i=1}^{n_j} X_{ij} \quad \text{and} \quad \overline{X}_{**} = \frac{1}{n} \sum_{j=1}^{k} \sum_{i=1}^{n_j} X_{ij}.$$

The quantity $\overline{X}_{*j}$ represents the sample mean of the jth sample, and $\overline{X}_{**}$ is the mean across all the samples.

By considering the identity $X_{ij} - \overline{X}_{**} = (X_{ij} - \overline{X}_{*j}) + (\overline{X}_{*j} - \overline{X}_{**})$, squaring both sides, and then taking the sum over i and j, we arrive at the following.

A Fundamental Identity

$$\underbrace{\sum_{j=1}^{k}\sum_{i=1}^{n_j}(X_{ij} - \overline{X}_{**})^2}_{SST} = \underbrace{\sum_{j=1}^{k}\sum_{i=1}^{n_j}(X_{ij} - \overline{X}_{*j})^2}_{SSE \text{ within treatments}} + \underbrace{\sum_{j=1}^{k}n_j(\overline{X}_{*j} - \overline{X}_{**})^2}_{SSTR \text{ between treatments}}$$

In the above equation, SST stands for **total sum-of-squares** and measures the total variation, SSE stands for **error sum-of-squares** and represents the sum of the variations within each sample, and finally, $SSTR$ stands for **treatment sum-of-squares** and represents the variation across samples. A considerable amount of algebra is required to verify the identity and is omitted.

The Test Statistic

Now suppose that the null hypothesis $H_0 : \mu_1 = \mu_2 = \cdots = \mu_k$ is **true**. In this case, $X_{11}, \ldots, X_{n_1 1}, \ldots, X_{1k}, \ldots, X_{n_k k}$ can be viewed as a random sample from $X \sim N(\mu, \sigma)$. Therefore, since sums of squares of independent standard normals is χ^2, we have

$$\frac{1}{\sigma^2}\sum_{j=1}^{k}\sum_{i=1}^{n_j}(X_{ij} - \overline{X}_{**})^2 = \frac{1}{\sigma^2}SST \sim \chi^2(n-1).$$

Similarly, for treatment n_j,

$$\frac{1}{\sigma^2}\sum_{i=1}^{n_j}(X_{ij} - \overline{X}_{*j})^2 \sim \chi^2(n_j - 1),$$

and so by independence,

$$\frac{1}{\sigma^2}\sum_{j=1}^{k}\sum_{i=1}^{n_j}(X_{ij} - \overline{X}_{*j})^2 = \frac{1}{\sigma^2}SSE \sim \chi^2(n-k)$$

since

$$\frac{1}{\sigma^2}SSE = \sum_{j=1}^{k}\left(\frac{1}{\sigma^2}\sum_{i=1}^{n_j}(X_{ij} - \overline{X}_{*j})^2\right) \sim \chi^2(n_1 - 1) + \cdots + \chi^2(n_k - 1) \sim \chi^2(n-k).$$

Notice that $(1/\sigma^2)SSE \sim \chi^2(n-k)$ is true **whether or not H_0 is true**. So we have the distributions of SST and SSE. To complete the picture, we need to know something about how $SSTR$ is distributed. The way to determine this is to use the theorem from Chapter 2 that if two random variables have the same moment generating function, then they have the same distribution. Since $(1/\sigma^2)SST \sim \chi^2(n-1)$ under H_0, its mgf is

$$M_{(1/\sigma^2)SST}(t) = (1 - 2t)^{-(n-1)/2}.$$

Also, since $(1/\sigma^2)SSE \sim \chi^2(n-k)$, we have $M_{(1/\sigma^2)SSE}(t) = (1-2t)^{-(n-k)/2}$. By independence, and since $(1/\sigma^2)SST = (1/\sigma^2)SSE + (1/\sigma^2)SSTR$,

$$M_{(1/\sigma^2)SST}(t) = M_{(1/\sigma^2)SSE}(t) \cdot M_{(1/\sigma^2)SSTR}(t).$$

Therefore,

$$M_{(1/\sigma^2)SSTR}(t) = \frac{M_{(1/\sigma^2)SST}(t)}{M_{(1/\sigma^2)SSE}(t)} = \frac{(1-2t)^{-\frac{n-1}{2}}}{(1-2t)^{-\frac{n-k}{2}}} = (1-2t)^{-(k-1)/2},$$

which is the mgf of a χ^2 random variable with $k - 1$ degrees of freedom. Therefore, under H_0,

$$\boxed{\frac{1}{\sigma^2} \sum_{j=1}^{k} n_j (\overline{X}_{*j} - \overline{X}_{**})^2 = (1/\sigma^2)SSTR \sim \chi^2(k-1).}$$

Now consider the expected values of SSE and $SSTR$. If H_0 is true, it is not hard to show that

$$E(SSE) = (n-k)\sigma^2 \text{ and } E(SSTR) = (k-1)\sigma^2.$$

In summary, assuming H_0, we have

$$\frac{1}{\sigma^2} SSE \sim \chi^2(n-k), \quad \frac{1}{\sigma^2} SSTR \sim \chi^2(k-1),$$
$$E(SSE) = (n-k)\sigma^2, \text{ and } E(SSTR) = (k-1)\sigma^2.$$

Therefore, since the ratio of χ^2 rvs, each divided by its respective degrees of freedom has an F-distribution as shown in Section 2.6.3, we have

$$\frac{\frac{1}{\sigma^2}SSTR}{k-1} \bigg/ \frac{\frac{1}{\sigma^2}SSE}{n-k} = \frac{SSTR/(k-1)}{SSE/(n-k)} \sim F(k-1, n-k). \tag{5.7}$$

If H_0 is true, since $E(SSE) = (n-k)\sigma^2$ and $E(SSTR) = (k-1)\sigma^2$, we would expect the ratio to be close to 1. However, if H_0 is **not** true, then it can be shown (details omitted) that

$$E(SSTR) = (k-1)\sigma^2 + \sum_{j=1}^{k} n_j(\mu_j - \mu)^2 > (k-1)\sigma^2$$

since the μ_js are not the same. The denominator in (5.7) should be about σ^2 since $E(SSE) = (n-k)\sigma^2$ even when H_0 is not true. The numerator, however, should be greater than σ^2, and so the ratio should be greater than 1. Therefore,

$$\boxed{F = \frac{SSTR/(k-1)}{SSE/(n-k)} \sim F(k-1, n-k)}$$

is our test statistic, and we will reject H_0 when its observed value, f, is too large. To prevent committing a Type I error more than $100\alpha\%$ of the time, we will **reject H_0 when $f \geq F(k - 1, n - k, \alpha)$**.

Remark 5.34   The calculations for ANOVA are simplified using the following formulas:

$$SST = \sum_{k=1}^{k} \sum_{i=1}^{n_j} X_{ij}^2 - \frac{1}{n} \left(\sum_{k=1}^{k} \sum_{i=1}^{n_j} X_{ij} \right)^2$$

$$SSTR = \sum_{j=1}^{k} \frac{S_j^2}{n_j} - \frac{1}{n} \left(\sum_{k=1}^{k} \sum_{i=1}^{n_j} X_{ij} \right)^2 \left(\text{where } S_j = \sum_{i=1}^{n_j} X_{ij} \right)$$

$$SSE = SST - SSTR.$$

To organize the calculations, ANOVA tables are used. Table entries are obtained by substituting data values for the X_{ij} in the formulas in the above remark.

Source	DF	SS	MSS	F-statistic	p-Value
Treatment	$k - 1$	$SSTR$	$MSTR = \frac{SSTR}{k-1}$	$f = \frac{MSTR}{MSE}$	$P(F(k - 1, n - k) \geq f)$
Error	$n - k$	SSE	$MSE = \frac{SSE}{n-k}$	*	*
Total	$n - 1$	*	*	*	*

The table is filled in with numerical values from left to right until a p-value is calculated.

Example 5.35   Sweetcorn, as opposed to field corn which is fed to livestock, is for human consumption. Three samples from three different types of sweetcorn, Sweetness, Allure, and Montauk, were compared to determine if the mean heights of the plants were the same. The following table gives the height (in feet) of 17 samples of Sweetness, 12 samples of Allure, and 15 samples of Montauk. Assume normal distributions for each treatment. Each treatment also has a variance of 0.64 feet.

Sweetness		Allure		Montauk	
5.48	6.06	7.51	6.73	5.83	5.42
5.21	6.14	6.54	5.85	5.80	6.92
5.08	4.99	6.66	7.28	6.27	5.41
4.14	5.88	5.29	6.83	5.44	6.65
6.56	5.49	5.17	6.77	6.54	5.60
4.81	6.81	7.45	7.25	6.78	6.05
6.70	6.44			7.19	6.08
6.99	6.66			6.16	
6.37					

We list the results of the ANOVA analysis in the following table.

Source	DF	SS	MSS	F-statistic	p-Value
Treatment	2	3.860	1.930	$f = 3.543$	0.038
Error	41	22.336	0.545	*	*
Total	43	*	*	*	*

If we set $\alpha = 0.05$, then $F(2, 41, 0.05) = 3.226 \leq f = 3.543$, and so we reject $H_0 : \mu_S = \mu_A = \mu_M$. The p-value is calculated from $P(F(2, 41) \geq 3.543) = F\text{cdf}(3.543, \infty, 2, 41) = 0.038 < \alpha$. The data is statistically significant evidence against the null, and we conclude that the mean heights of the three types of plants are not the same.

Remark 5.36   To calculate the value $F(d1, d2, \alpha)$ for a given $\alpha, d1$, and $d2$, using a TI calculator, enter the following program as INVF into your calculator:

(a) Input "RT TAIL," A

(b) Input "D1:," N

(c) Input "D2:," D

(d) solve(1-Fcdf(0,X,N,D)-A, X, 1.5*N,0,9999) → X

(e) Disp X

(f) Stop

Press PRGM, then INVF, and you will be prompted for the area to the right of c, i.e., α as in $P(F \geq c) = \alpha$. Then enter the degrees of freedom in the order D1, D2, and press ENTER. The result is the value of c that gives you α.

The Special Case of Two Samples

Let $X_{11}, X_{21}, \ldots, X_{n_1 1}$ and $X_{12}, X_{22}, \ldots, X_{n_2 2}$ be independent random samples from random variables $X_1 \sim N(\mu_1, \sigma)$ and $X_2 \sim N(\mu_2, \sigma)$, respectively. (Note that the variances are equal.)

There are two methods we can use to test the hypotheses

$$H_0 : \mu_1 = \mu_2 \quad \text{vs.} \quad H_1 : \mu_1 \neq \mu_2.$$

One test is the two-sample t-test described earlier in this chapter which uses the statistic

$$T = \frac{(\overline{X}_{*1} - \overline{X}_{*2})}{S_p \sqrt{\frac{1}{n_1} + \frac{1}{n_2}}} \sim t(n_1 + n_2 - 2)$$

where S_p^2 is the pooled variance. Or we could use the ANOVA F-test just introduced. Which of these methods is better and in what sense? To answer this question, we will apply the ANOVA procedure to the two-sample problem. In this situation, the F statistic becomes

$$F = \frac{MSTR}{MSE} = \frac{SSTR/(k-1)}{SSE/(n-k)} = \frac{SSTR}{SSE/(n_1 + n_2 - 2)}$$

since $k = 2$ and $n = n_1 + n_2$. Computing $SSTR$, we get

$$SSTR = n_1 \left(\overline{X}_{*1} - \frac{n_1 \overline{X}_{*1} + n_2 \overline{X}_{*2}}{n_1 + n_2} \right)^2 + n_2 \left(\overline{X}_{*2} - \frac{n_1 \overline{X}_{*1} + n_2 \overline{X}_{*2}}{n_1 + n_2} \right)^2$$

$$= n_1 \left(\frac{n_2 \left(\overline{X}_{*1} - \overline{X}_{*2} \right)}{n_1 + n_2} \right)^2 + n_2 \left(\frac{n_1 \left(\overline{X}_{*1} - \overline{X}_{*2} \right)}{n_1 + n_2} \right)^2$$

$$= \frac{n_1 n_2^2 + n_2 n_1^2}{(n_1 + n_2)^2} \left(\overline{X}_{*1} - \overline{X}_{*2} \right)^2$$

$$= \frac{n_1 n_2}{n_1 + n_2} \left(\overline{X}_{*1} - \overline{X}_{*2} \right)^2 = \frac{\left(\overline{X}_{*1} - \overline{X}_{*2} \right)^2}{\frac{1}{n_1} + \frac{1}{n_2}}.$$

Computing SSE, we obtain

$$SSE = \sum_{j=1}^{2} \sum_{i=1}^{n_j} (X_{ij} - \overline{X}_{*j})^2 = (n_1 - 1)S_{X_1}^2 + (n_2 - 1)S_{X_2}^2,$$

and so

$$\frac{SSE}{n_1 + n_2 - 2} = \frac{(n_1 - 1)S_{X_1}^2 + (n_2 - 1)S_{X_2}^2}{n_1 + n_2 - 2},$$

which is just the pooled variance S_p^2. Therefore,

$$\frac{SSTR}{SSE/(n_1 + n_2 - 2)} = \frac{\left(\overline{X}_{*1} - \overline{X}_{*2} \right)^2}{S_p^2 \left(\frac{1}{n_1} + \frac{1}{n_2} \right)} \sim F(1, n_1 + n_2 - 2).$$

We reject $H_0 : \mu_1 = \mu_2$ at level α if

$$f = \frac{(\overline{x}_{*1} - \overline{x}_{*2})^2}{s_p^2 \left(\dfrac{1}{n_1} + \dfrac{1}{n_2}\right)} \geq F(1, n_1 + n_2 - 2, \alpha).$$

However, we know that $F(1, m) = (t(m))^2$, and so we reject H_0 when

$$\frac{(\overline{x}_{*1} - \overline{x}_{*2})^2}{\left(\dfrac{1}{n_1} + \dfrac{1}{n_2}\right) s_p^2} \geq (t(n_1 + n_2 - 2, \alpha))^2 \iff |t| \geq t(n_1 + n_2 - 2, \alpha)$$

where t is the value of the statistic

$$T = \frac{\overline{X}_{*1} - \overline{X}_{*2}}{S_p \sqrt{\dfrac{1}{n_1} + \dfrac{1}{n_2}}}.$$

This is *exactly* the condition under which we reject H_0 using the T statistic for a two-sample t-test. The two methods are equivalent in that one of the methods will reject H_0 if and only if the other method does.

5.7   PROBLEMS

5.1. An industrial drill bit has a lifetime (measured in years) that is a normal random variable $X \sim N(\mu, 2)$. A random sample of the lifetimes of 100 bits resulted in a sample mean of $\overline{x} = 1.3$.

(a) Perform a test of hypothesis $H_0 : \mu_0 = 1.5$ vs. $H_1 : \mu_0 \neq 1.5$ with $\alpha = 0.05$.

(b) Compute the probability of a Type II error when $\mu = 2$. That is, compute $\beta(2)$.

(c) Find a general expression for $\beta(\mu)$.

5.2. Suppose a test of hypothesis is conducted by an experimenter for the mean of a normal random variable $X \sim N(\mu, 3.2)$. A random sample of size 16 is taken from X. If $H_0 : \mu = 42.9$ is tested against $H_1 : \mu \neq 42.9$, and the experimenter rejects H_0 if $\overline{x}$ is in the region $(-\infty, 41.164] \cup [44.636, \infty)$, what is the level of the test α?

5.3. Compute the p-values associated with each of the following sample means $\overline{x}$ computed from a random sample from a normal random variable X. Then decide if the null hypothesis should be rejected if $\alpha = 0.05$.

(a) $H_0 : \mu = 120$ vs. $H_1 : \mu < 120$, $n = 25$, $\sigma = 18$, $\overline{x} = 114.2$.

(b) $H_0 : \mu = 14.2$ vs. $H_1 : \mu > 14.2$, $n = 9$, $\sigma = 4.1$, $\overline{x} = 15.8$.

(c) $H_0 : \mu = 30$ vs. $H_1 : \mu \neq 30$, $n = 16$, $\sigma = 6$, $\bar{x} = 26.8$.

5.4. An engineer at a prominent manufacturer of high performance alkaline batteries is attempting to increase the lifetime of its best-selling AA battery. The company's current battery functions for 100.3 hours before it has to be recharged. The engineer randomly selects 15 of the improved batteries and discovers that the mean operating time is $\bar{x} = 105.6$ hours. The sample standard deviation is $s = 6.25$.

(a) Perform a test of hypotheses
$$H_0 : \mu = 100.3 \text{ vs. } H_1 : \mu > 100.3 \text{ with } \alpha = 0.01.$$

(b) Compute the power function value $\pi(103)$.

(c) Compute the general form of the power function $\pi(\mu)$.

5.5. Suppose a random sample of size 25 is taken from a normal random variable $X \sim N(\mu, \sigma)$ with σ unknown. For the following tests and test levels, determine the critical regions if t represents the value of the test statistic $T = \dfrac{\bar{X} - \mu_0}{s_X/5}$.

(a) $H_0 : \mu > \mu_0$ vs. $H_1 : \mu = \mu_0$, $\alpha = 0.01$.

(b) $H_0 : \mu < \mu_0$ vs. $H_1 : \mu = \mu_0$, $\alpha = 0.02$.

(c) $H_0 : \mu = \mu_0$ vs. $H_1 : \mu \neq \mu_0$, $\alpha = 0.05$.

5.6. Compute the p-values associated with each of the following sample means $\bar{x}$ computed from a random sample from a normal random variable $X \sim N(\mu, \sigma)$ where σ is unknown. Then decide if the null hypothesis should be rejected if $\alpha = 0.05$.

(a) $H_0 : \mu = 90.5$ vs. $H_1 : \mu < 90.5$, $s = 9.5$, $n = 17$, $\bar{x} = 85.2$.

(b) $H_0 : \mu = 20.2$ vs. $H_1 : \mu > 20.2$, $s = 6.3$, $n = 9$, $\bar{x} = 21.8$.

(c) $H_0 : \mu = 35$ vs. $H_1 : \mu \neq 35$, $s = 11.7$, $n = 20$, $\bar{x} = 31.9$.

5.7. Consider the calculation of a Type II error for a test for the variance of a random variable $X \sim N(\mu, \sigma)$. Derive the general forms of $\beta(\sigma_1^2)$ for the following tests.

(a) $H_0 : \sigma^2 = \sigma_0^2$ vs. $H_1 : \sigma^2 < \sigma_0^2$ has $\beta(\sigma_1^2)$
$$= P\left(\chi^2(n-1) > \frac{\sigma_0^2}{\sigma_1^2} \chi^2(n-1, 1-\alpha) \right).$$

(b) $H_0 : \sigma^2 = \sigma_0^2$ vs. $H_1 : \sigma^2 > \sigma_0^2$ has $\beta(\sigma_1^2)$
$$= P\left(\chi^2(n-1) < \frac{\sigma_0^2}{\sigma_1^2} \chi^2(n-1, \alpha) \right).$$

(c) $H_0 : \sigma^2 = \sigma_0^2$ vs. $H_1 : \sigma^2 \neq \sigma_0^2$ has $\beta(\sigma_1^2)$
$$= P\left(\frac{\sigma_0^2}{\sigma_1^2} \chi^2(n-1, 1-\alpha/2) < \chi^2(n-1) < \frac{\sigma_0^2}{\sigma_1^2} \chi^2(n-1, \alpha/2) \right).$$

5.8. Pet Friendly, a pet supply company, sells 25-lb bags of a popular brand of cat litter, OdorGone. When properly filled, the bags have a standard deviation of 1 lb of litter. A random sample of 20 bags of litter are weighed with the results listed below.

25.23   25.50   26.18   25.44   26.04   26.01   25.30   24.49   25.21   25.68
25.18   25.01   26.09   24.49   24.54   25.12   25.84   24.22   25.14   25.67

(a) Perform a test of hypothesis $H_0 : \sigma^2 = 1$ vs. $H_1 : \sigma^2 > 1$ with $\alpha = 0.05$.

(b) Compute the probability of a Type II error when $\sigma^2 = 1.5$, and then compute $\pi(1.5)$.

(c) Determine the general form of $\pi(\sigma^2)$.

5.9. Compute the p-values associated with each of the following sample variances s_X^2 computed from a random sample from a normal random variable $X \sim N(\mu, \sigma)$. Then decide if the null hypothesis should be rejected if $\alpha = 0.05$.

(a) $H_0 : \sigma^2 = 4.8$ vs. $H_1 : \sigma^2 < 4.8$, $s_X^2 = 3.4$, $n = 60$.

(b) $H_0 : \sigma^2 = 2.3$ vs. $H_1 : \sigma^2 > 2.3$, $s_X^2 = 3.1$, $n = 35$.

(c) $H_0 : \sigma = 9$ vs. $H_1 : \sigma < 9$, $s_X = 7.1$, $n = 20$.

5.10. An electrical supply company produces devices that operate using a thermostatic control. The standard deviation of the temperature at which these devices actually operate should not exceed 2°C. The quality control department tests $H_0 : \sigma = 2$ vs. $H_1 : \sigma > 2$. A random sample of 30 devices is taken, and it is determined that $s = 2.39°C$.

(a) Conduct the test with $\alpha = 0.05$.

(b) Compute the p-value of the test.

(c) Compute the probability of a Type II error when $\sigma = 3$. Compute $\pi(3)$.

5.11. The personnel department at a suburban Chicago skilled nursing facility wants to test the hypothesis that $\sigma = 2$ hours for the time it takes a nurse to complete his or her round of patients against the alternative hypothesis that $\sigma \neq 2$. A random sample of the times required by 30 nurses to complete their rounds resulted in a sample standard deviation of $s = 1.8$ hours. Perform a test of hypothesis $H_0 : \sigma = 2$ vs. $H_1 : \sigma \neq 2$ with $\alpha = 0.05$.

5.12. The production of plastic sheets used in the construction industry is monitored at a company's plant for possible fluctuations in thickness (measured in millimeters, mm). If the variance in the thickness of the sheets exceeds 2.25 square millimeters, there is cause for concern about product quality. The production process continues while the variance appears to be smaller than the cutoff. Thickness measurements for a simple random sample of 10 sheets produced during a particular shift were taken, yielding the following result.

$$226, 226, 227, 226, 225, 228, 225, 226, 229, 227$$

(a) Perform a test of hypotheses $H_0 : \sigma^2 = 2.25$ vs. $H_1 : \sigma^2 < 2.25$ with $\alpha = 0.05$.

(b) Calculate the Type II error $\beta(2)$.

5.13. Cars intending on making a left turn at a certain intersection are observed. Out of 600 cars in the study, 157 of them pulled into the wrong lane.

(a) Perform a test of hypothesis $H_0 : p_0 = 0.30$ vs. $H_1 : p_0 \neq 0.30$ with $\alpha = 0.05$ where p_0 is the true proportion of cars pulling into the wrong left turn lane.

(b) Compute the p-value of the test.

(c) Compute $\beta(0.27)$.

5.14. A die is tossed 800 times. Consider the hypothesis $H_0 : p_0 = \frac{1}{6}$ vs. $H_1 : p_0 \neq \frac{1}{6}$ where p_0 is the proportion of 6s that appear.

(a) What is the range on the number of times x that a 6 would have to be rolled to reject H_0 at the $\alpha = 0.05$ level?

(b) What is the range on the number of times x that a 6 would have to be rolled to retain H_0 at the $\alpha = 0.01$ level?

5.15. (Small Sample Size) A cosmetics firm claims that its new topical cream reduces the appearance of undereye bags in 60% of men and women. A consumer group thinks the percentage is too high and conducts a test of hypothesis $H_0 : p_0 = 0.6$ vs. $H_1 : p_0 < 0.6$. Out of 8 men and women in a random sample, only 3 saw a significant reduction in the appearance of their undereye bags.

(a) Perform a test of hypothesis $H_0 : p_0 = 0.6$ vs. $H_1 : p_0 < 0.6$ with $\alpha = 0.05$ by computing the p-value of the test.

(b) What is the critical region of the test $H_0 : p_0 = 0.6$ vs. $H_1 : p_0 \neq 0.6$ if $\alpha = 0.1$?

5.16. The director of a certain university tutoring center wants to know if there is a difference between the mean lengths of time (in hours) male and female freshman students study over a 30-day period. The study involved a random sample of 34 female and 29 male students. Sample means and standard deviations are listed in the table below.

Female (X)	$m = 34$	$\bar{x} = 105.5$	$s_X = 20.1$
Male (Y)	$n = 29$	$\bar{y} = 90.9$	$s_Y = 12.2$

(a) Can it be assumed that $\sigma_X^2 = \sigma_Y^2$? Test the hypothesis $H_0 : \sigma_X^2 = \sigma_Y^2$ vs. $H_1 : \sigma_X^2 \neq \sigma_Y^2$ at the $\alpha = 0.05$ level.

(b) Depending on the outcome of the test in (a), test the hypothesis $H_0 : \mu_X = \mu_Y$ vs. $H_1 : \mu_X \neq \mu_Y$ again at the $\alpha = 0.05$ level.

(c) What is the p-value of the test in (b)?

5.17. Random samples of sizes $m = 11$ and $n = 10$ are taken from two independent normal random variables X and Y, respectively. The samples yield $s_X^2 = 6.8$ and $s_Y^2 = 7.1$. Perform the following tests.

(a) $H_0 : \sigma_X^2 = \sigma_Y^2$ vs. $H_1 : \sigma_X^2 > \sigma_Y^2$ with $\alpha = 0.1$.

(b) $H_0 : \sigma_X^2 = \sigma_Y^2$ vs. $H_1 : \sigma_X^2 < \sigma_Y^2$ with $\alpha = 0.05$.

(c) $H_0 : \sigma_X^2 = \sigma_Y^2$ vs. $H_1 : \sigma_X^2 \neq \sigma_Y^2$ with $\alpha = 0.01$.

5.18. A zookeeper at a major U.S. zoo wants to know if polar bears kept in captivity have a lower birth rate than those in the wild. The peak reproductive years are between 15 and 30 years of age in both captive and wild bears. Random samples of the number of cubs born to female captive bears and wild bears (from a certain Arctic population) were taken with the following results.

Captive Bears (X)	Wild Bears (Y)
$m = 24$	$n = 18$
$\overline{x} = 19.1$	$\overline{y} = 16.3$
$s_X = 2.3$	$s_Y = 4.1$

The variances of the two populations are unknown but assumed equal.

(a) Perform a test of hypothesis $H_0 : \mu_X = \mu_Y$ vs. $H_1 : \mu_X > \mu_Y$ with $\alpha = 0.05$.

(b) Compute the p-value of test in (a).

5.19. An entomologist is making a study of two species of lady bugs (Coccinellidae). She is interested in whether there is a difference between the number of spots on the carapace of the two species. She takes a random sample of 20 insects from each species and counts the number of spots. The results are presented in the table below.

Species 1 (X)	Species 2 (Y)
$m = 20$	$n = 20$
$\overline{x} = 3.8$	$\overline{y} = 3.6$
$s_X = 1.2$	$s_Y = 1.3$

(a) Perform a test of hypothesis $H_0 : \mu_X = \mu_Y$ vs. $H_1 : \mu_X \neq \mu_Y$ with $\alpha = 0.05$.

(b) Compute the p-value of test in (a).

5.20. Random samples of sizes $m = 9$ and $n = 14$ are taken from two independent normal random variables X and Y, respectively. The variances are unknown but assumed equal. Assume that the pooled variance $s_p^2 = 3581.6$. If $H_0 : \mu_X = \mu_Y$ vs. $H_1 : \mu_X \neq \mu_Y$ is to be tested with $\alpha = 0.05$, what is the smallest value of $|\overline{x} - \overline{y}|$ that will result in the null hypothesis being rejected?

5.21. A nutritionist wishes to study the effects of diet on heart disease and stroke in middle-aged men. He conducts a study of 1,000 randomly selected initially healthy men between the ages of 45 and 60. Exactly half of the men were placed on a restricted diet (X sample) while the other half were allowed to continue their normal diet (Y sample). After an 8-year period, 85 men in the diet group had died of myocardial infarction (heart attack) or cerebral infarction (stroke) while 93 men died of heart attack or stroke in the control group.

(a) Perform a test of hypothesis $H_0 : p_X = p_Y$ vs. $H_1 : p_X \neq p_Y$ with $\alpha = 0.05$.

(b) What is the p-value of the test in (a)?

(c) Calculate $\beta(-.05)$.

5.22. A utility outfielder on the Chicago Cubs had a batting average of 0.276 out of 300 at bats last season and a batting average of 0.220 out of 235 at bats this past season. Management wants to reduce the salary of the player for next season since his performance appears to have degraded. Does management have a sound statistical argument for cutting the player's salary?

(a) Perform a suitable test of hypothesis to either validate or invalidate management's decision.

(b) Compute the p-value of the test in (a).

5.23. A random sample of 550 Californians (X sample) and 690 Iowans (Y sample) were asked if they would like to visit Europe within the next ten years with the result that 61% of Californians and 53% of Iowans would like to take the trip.

(a) Perform a test of hypothesis $H_0 : p_X = p_Y$ vs. $H_1 : p_X > p_Y$ with $\alpha = 0.01$.

(b) Compute the p-value of the test in (a).

(c) Compute $\beta(0.1)$.

5.24. Ten students at a certain high school are randomly chosen to participate in a study of the effectiveness of a taking a course in formal logic on their abstract reasoning capability. The students take a test measuring abstract reasoning before and after taking the course. The results of the two tests are displayed in the following table.

Student i	Score Before Course X_i	Score After Course Y_i	Difference $D = Y_i - X_i$
1	74	78	4.0
2	83	79	−4.0
3	75	76	1.0
4	88	85	−3.0
5	84	86	2.0
6	63	67	4.0
7	93	93	0.0
8	84	83	−1.0
9	91	94	3.0
10	77	76	−1.0

(a) Perform a paired t−test of hypothesis $H_0 : \mu_D = 0$ vs. $H_1 : \mu_D > 0$ with $\alpha = 0.05$.

(b) What is the p-value of the test in (a)?

5.25. A new diet requires that certain food items be weighed before being consumed. Over the course of a week, a person on the diet weighs ten food items (in ounces). Just to make sure of the weight, she weighs the items on two different scales. The weights indicated on the scales are close to one another, but are not exactly the same. The results of the weighings are given below.

Food Item i	Weight X_i on Scale 1	Weight Y_i on Scale 2	Difference $D_i = X_i - Y_i$
1	19.38	19.35	0.03
2	12.40	12.45	−0.05
3	6.47	6.46	0.01
4	13.47	13.52	−0.05
5	11.23	11.27	−0.04
6	14.36	14.41	−0.05
7	8.33	8.35	−0.02
8	10.50	10.52	−0.02
9	23.42	23.41	0.01
10	9.15	9.17	−0.02

(a) Perform a paired t−test of hypothesis $H_0 : \mu_D = 0$ vs. $H_1 : \mu_D \neq 0$ with $\alpha = 0.05$.

(b) What is the p-value of the test in (a)?

5.26. The table on the left lists the percentages in the population of each blood type in India. The table on the right is the distribution of 1,150 blood types in a small northern Indian town.

Blood Type	Percentage of Population
O+	27.85
A+	20.8
B+	38.14
AB+	8.93
O-	1.43
A-	0.57
B-	1.79
AB-	0.49

Blood Type	Number of Residents
O+	334
A+	207
B+	448
AB+	92
O-	23
A-	12
B-	23
AB-	11

Does the town's distribution of blood types conform to the national percentages for India? Test at the $\alpha = 0.01$ level.

5.27. A six-sided die is tossed independently 180 times. The following frequencies were observed.

Side	1	2	3	4	5	6
Frequency	ϑ	30	30	30	30	$60 - \vartheta$

For what values of ϑ would the null hypothesis that the die is fair be rejected at the $\alpha = 0.05$ level?

5.28. The distribution of colors in the candy M&Ms has varied over the years. A statistics student conducted a study of the color distribution of M&Ms made in a factory in Tennessee. After the study, she settles on the following distribution of the colors blue, orange, green, yellow, red, and brown at the Tennessee plant.

Color	Percentage (Tennessee)
Blue	20.7
Orange	20.5
Green	19.8
Yellow	13.5
Red	13.1
Brown	12.4

Color	Number (New Jersey)
Blue	213
Orange	208
Green	183
Yellow	131
Red	137
Brown	128

She wanted to see if the same distribution held at another plant located in New Jersey. A sample of 1,000 M&Ms were inspected for color at the New Jersey plant. The results are in the table above right. Is the New Jersey plant's distribution of colors the same as the Tennessee plant's distribution? Test at the $\alpha = 0.05$ level.

5.29. At a certain fishing resort off the Southeastern cost of Florida, a record is kept of the number of sailfish caught daily over a 60-day period by the guests staying at the resort.

The results are in the table below.

Sailfish Caught	0	1	2	3	4	5	≥ 6
No. Days	8	14	14	17	3	3	1

An ecologist is concerned about declining fish populations in the area. He proposes that the data follows a Poisson distribution with a Poisson rate $\lambda = 2$. Does it appear that the data follows such a distribution? Test at the $\alpha = 0.05$ level.

5.30. A homeowner is interested in attracting hummingbirds to her backyard by installing a bird feeder customized for hummingbirds. Over a period of 576 days, the homeowner observes the number of hummingbirds visiting the feeder during a certain half-hour period during the afternoon.

Hummingbird Visits	0	1	2	3	4	≥ 5
No. Days	229	211	93	35	7	1

(a) Is it possible the data in the table follows a Poisson distribution? Test at the $\alpha = 0.05$ level. What value of λ should be used?

(b) Apply the test with $\lambda = 0.8$

5.31. A traffic control officer is tracking speeders on Lake Shore Drive in Chicago. In particular, he is recording (from a specific vantage point) the time intervals (interarrival times) between drivers that are speeding. A sample of 100 times (in seconds) are listed in the table below.

Interval	$[0, 20)$	$[20, 40)$	$[40, 60)$	$[60, 90)$	$[90, 120)$	$[120, 180)$	$[180, \infty)$
No. in the Interval	41	19	16	13	9	2	0

The officer hypothesizes that the interarrival time data follows an exponential distribution with mean 40 seconds. Test his hypothesis at the $\alpha = 0.05$ level.

5.32. A fisherman, who also happens to be a statistician, is counting the number of casts he has to make in a certain small lake in southern Illinois near his home before his lure is taken by a smallmouth bass. The data below represents the number of casts until achieving 50 strikes while fishing during a recent vacation. The fisherman hypothesizes that the number of casts before a strike follows a geometric distribution. Test his hypothesis at the $\alpha = 0.01$ level.

Casts to Strike	1	2	3	4	5	6	7	8	9
Frequency	4	13	10	7	5	4	3	3	1

5.33. A criminologist is studying the occurrence of serious injury due to criminal violence for certain professions. A random sample of 490 causes of injury in the chosen professions is taken. The results are displayed in the table. Does it appear that serious injury due to criminal violence and choice of profession are independent? Test at the $\alpha = 0.01$ level.

	Profession				
	Police (P)	Cashier (C)	Taxi Driver (T)	Security Guard (S)	Row Totals
Criminal Violence (V)	82	107	70	59	318
Other Causes (O)	92	9	29	42	172
Column Totals	174	116	99	101	490

5.34. For a receiver in a wireless device, like a cell phone, for example, two important characteristics are its selectivity and sensitivity. Selectivity refers to a wireless receiver's capability to detect and decode a desired signal in the presence of other unwanted interfering signals. Sensitivity refers to the smallest possible signal power level at the input which assures proper functioning of a wireless receiver. A random sample of 170 radio receivers produced the following results.

		Sensitivity			
		Low (LN)	Average (AN)	High (HN)	Row Totals
	Low (LS)	6	12	12	30
Selectivity	Average (AS)	33	61	18	112
	High (HS)	13	15	0	28
	Column Totals	52	88	30	170

Does it appear from the data that selectivity and sensitivity are dependent traits of a receiver? Test at the $\alpha = 0.01$ level.

5.35. Consider again the small northern Indian town with 1,150 residents in problem 5.26 that has the following distribution of blood types.

Blood Type	O+	A+	B+	AB+	O-	A-	B-	AB-
No.Residents	334	207	448	92	23	12	23	11

Test the hypothesis that blood type and rH factor (positive or negative) are independent traits at the $\alpha = 0.05$ level.

5.36. A random sample of 300 adults in a certain small city in Illinois are asked about their favorite PBS programs. In particular, they are asked about their favorite shows among home improvement shows, cooking shows, and travel shows. The results of the survey are listed in the table below.

		PBS Program			
		Home Improvement (H)	Cooking (C)	Travel (T)	Row Totals
Gender	Male (M)	45	65	50	160
	Female (F)	55	45	40	140
	Column Totals	100	110	90	300

Is gender and the genre of television program independent? Test the hypothesis at the $\alpha = 0.05$ level.

5.37. A ketogenic diet is a type of low-carbohydrate diet whose aim is to metabolize fats into ketone bodies (water-soluble molecules acetoacetate, beta-hydroxybutyrate, and acetone produced by the liver) rather than into glucose as the body's main source of energy. A dietician wishes to study whether the proportion of adult men on ketogenic diets changes with age. She samples 100 men currently on diets in each of five age groups: Group I: 20–25, Group II: 26–30, Group III: 31–35, Group IV: 36–40, and Group V: 41–45. Her results are displayed in the table below. Let $p_i, i \in \{I, II, III, IV, V\}$ denote the proportion of men on a ketogenic diet in group i. Test whether the proportions are the same across the age groups. Conduct the test at the $\alpha = 0.05$ level.

		Group					
		(I)	(II)	(III)	(IV)	(V)	Row Totals
Diet	Ketogenic (K)	26	22	25	20	19	112
	Nonketogenic (N)	74	78	75	80	81	388
	Column Totals	100	100	100	100	100	500

5.38. Suppose the result of an experiment can be classified as having one of three mutually exclusive A traits, A_1, A_2, and A_3, and also as having one of four mutually exclusive B traits, B_1, B_2, B_3, and B_4. The experiment is independently repeated 300 times with the following results.

		B Trait				
		B_1	B_2	B_3	B_4	Row Totals
A Trait	A_1	$25 - 5\vartheta$	$25 - \vartheta$	$25 + \vartheta$	$25 + 5\vartheta$	100
	A_2	25	25	25	25	100
	A_3	$25 + 5\vartheta$	$25 + \vartheta$	$25 - \vartheta$	$25 - 5\vartheta$	100
	Column Totals	75	75	75	75	300

What is the smallest integer value of ϑ for which the null hypothesis that the traits are independent is rejected? Test at the $\alpha = 0.05$ level. (Note that $0 \leq \vartheta \leq 5$.)

5.39. A random sample of size $n = 10$ of top speeds of Indianapolis 500 drivers over the past 30 years is taken. The data is displayed in the following table.

Top Speeds (in mph)				
202.2	203.4	200.5	206.3	198.0
203.7	200.8	201.3	199.0	202.5

Test the set of hypotheses

$$H_0 : m_0 = 200 \quad \text{vs.} \quad H_1 : m_0 > 200$$

at the $\alpha = 0.05$ level.

5.40. Random samples of size 6 are taken from three normal random variables A, B, and C having equal variances. The results are displayed in the table below.

Variable A (A)	Variable B (B)	Variable C (C)
15.75	12.63	9.37
11.55	11.46	8.28
11.16	10.77	8.15
9.92	9.93	6.37
9.23	9.87	6.37
8.20	9.42	5.66

Test if $\mu_A = \mu_B = \mu_C$ at the $\alpha = 0.01$ level by carrying out the following steps.

(a) Compute $\overline{X}_{*A}$, $\overline{X}_{*B}$, $\overline{X}_{*C}$, and $\overline{X}_{**}$.

(b) Compute $SSTR$, SSE, and SST.

(c) Compute $MSTR$ and MSE.

(d) Compute f.

(e) Compute the p-value of f.

(f) Display the ANOVA table.

(g) Test if $\mu_A = \mu_B = \mu_C$ at the $\alpha = 0.01$ level.

5.41. Alice, John, and Bob are three truck assembly plant workers in Dearborn, MI. The times (in minutes) each requires to mount the windshield on a particular model of truck the plant produces are recorded on five randomly chosen occasions for each worker.

Alice (A)	John (J)	Bob (B)
8	10	11
10	8	10
10	9	10
11	9	9
9	8	9

Assuming equal variances, test if $\mu_A = \mu_J = \mu_B$ at the $\alpha = 0.05$ level by carrying out the following steps.

(a) Compute $\overline{X}_{*A}$, $\overline{X}_{*J}$, $\overline{X}_{*B}$, and $\overline{X}_{**}$.

(b) Compute $SSTR$, SSE, and SST.

(c) Compute $MSTR$ and MSE.

(d) Compute f.

(e) Compute the p-value of f.

(f) Display the ANOVA table.

(g) Test if $\mu_A = \mu_J = \mu_B$ at the $\alpha = 0.05$ level.

5.42. A new drug, AdolLoft, was developed to treat depression in adolescents. A research study was established to assess the clinical efficacy of the drug. Patients suffering from depression were randomly assigned to one of three groups: a placebo group (P), a low-dose group (L), and a normal dose group (N). After six weeks, the subjects completed the Beck Depression Inventory (BDI-II, 1996) assessment which is composed of questions relating to symptoms of depression such as hopelessness and irritability, feelings of guilt or of being punished, and physical symptoms such as tiredness and weight loss. The results of the study on the three groups of five subjects is given below.

Placebo (P)	Low Dose (L)	Normal Dose (N)
38	31	11
42	c23	5
25	22	26
47	19	18
39	8	14

For the BDI-II assessment, a score of 0–13 indicates minimal depression, 14–19 indicates mild depression, 20–28 indicates moderate depression, and 29–63 indicates severe depression. Assuming identical variances, test if $\mu_P = \mu_L = \mu_N$ at the $\alpha = 0.01$ level.

5.43. Milk from dairies located in central Illinois is tested for Strontium-90 contamination. The dairies are located in three counties: Macon (5 dairies), Sangamon (6 dairies), and

Logan (5 dairies). Contamination is measured in picocuries/liter.

Macon (M)	Sangamon (S)	Logan (L)
11.7	8.8	12.1
10.4	9.9	9.5
9.5	11.2	9.0
13.8	10.5	10.3
12.0	9.1	8, 7
	8.5	

Assuming equal variances, test if $\mu_M = \mu_S = \mu_L$ at the $\alpha = 0.05$ level.

5.44. Consider the following partially completed ANOVA table.

Source	DF	SS	MSS	F-statistic	p-Value
Treatment	____	2.124	0.708	0.75	____
Error	20	____	____	*	*
Total	____	____	*	*	*

Fill in the missing values.

5.45. The police department of a Chicago metropolitan suburb is conducting a study of marksmanship skill among police officers that have served varying lengths of time in the department. Groups of randomly selected officers that have served 5 years, 10 years, 15 years, and 20 years in the department are selected for the study. There are five officers in each group. Each officer is given 75 shots at a target at a distance of 25 yards, and the number of bull's-eyes are recorded.

5 Years (5Y)	10 Years (10Y)	15 Years (15Y)	20 Years (20Y)
60	55	45	42
59	57	39	41
62	64	43	39
56	61	45	43
55	50	46	45

Assuming equal variances, test if $\mu_{5Y} = \mu_{10Y} = \mu_{15Y} = \mu_{20Y}$ at the $\alpha = 0.01$ level by carrying out the following steps.

(a) Compute $\overline{X}_{*5Y}$, $\overline{X}_{*10Y}$, $\overline{X}_{*15Y}$, $\overline{X}_{*20Y}$, $\overline{X}_{**}$.

(b) Compute $SSTR$, SSE, and SST.

(c) Compute $MSTR$ and MSE.

(d) Compute f.

(e) Compute the p-value of f.

(f) Display the ANOVA table.

(g) Test if $\mu_{5Y} = \mu_{10Y} = \mu_{15Y} = \mu_{20Y}$ at the $\alpha = 0.01$ level.

5.8   SUMMARY TABLES

The following tables summarize the tests of hypotheses developed in this chapter.

Parameter	Conditions	Test	Decision Rule
μ	$\sigma = \sigma_0$ known test statistic: $Z = \dfrac{(\bar{X} - \mu_0)}{\sigma_0 / \sqrt{n}}$	$H_0 : \mu = \mu_0$ $H_1 : \mu > \mu_0$	Reject H_0 if $z \geq z_\alpha$
		$H_0 : \mu = \mu_0$ $H_1 : \mu < \mu_0$	Reject H_0 if $z \leq -z_\alpha$
μ	σ unknown test statistic: $T = \dfrac{(\bar{X} - \mu_0)}{S_X / \sqrt{n}}$	$H_0 : \mu = \mu_0$ $H_1 : \mu > \mu_0$	Reject H_0 if $t \geq t(n-1, \alpha)$
		$H_0 : \mu = \mu_0$ $H_1 : \mu < \mu_0$	Reject H_0 if $t \leq -t(n-1, \alpha)$
σ^2	$\mu = \mu_0$ known test statistic: $X^2 = \displaystyle\sum_{i=1}^{n} \left(\dfrac{X_i - \mu_0}{\sigma_0} \right)^2$	$H_0 : \sigma^2 = \sigma_0^2$ $H_1 : \sigma^2 > \sigma_0^2$	Reject H_0 if $\chi^2 \geq \chi^2(n, \alpha)$
		$H_0 : \sigma^2 = \sigma_0^2$ $H_1 : \sigma^2 < \sigma_0^2$	Reject H_0 if $\chi^2 \leq \chi^2(n, 1-\alpha)$
σ^2	μ unknown test statistic: $X^2 = \displaystyle\sum_{i=1}^{n} \left(\dfrac{X_i - \bar{X}_n}{\sigma_0} \right)^2$ $= \dfrac{(n-1)S_X^2}{\sigma_0^2}$	$H_0 : \sigma^2 = \sigma_0^2$ $H_1 : \sigma^2 > \sigma_0^2$	Reject H_0 if $\chi^2 \geq \chi^2(n-1, \alpha)$
		$H_0 : \sigma^2 = \sigma_0^2$ $H_1 : \sigma^2 < \sigma_0^2$	Reject H_0 if $\chi^2 \leq \chi^2(n-1, 1-\alpha)$

One-sided tests for normal parameters

Parameter	Conditions	Test	Decision Rule
μ	$\sigma = \sigma_0$ known test statistic: $Z = \dfrac{(\bar{X} - \mu_0)}{\sigma_0/\sqrt{n}}$	$H_0 : \mu = \mu_0$ $H_1 : \mu \neq \mu_0$	Reject H_0 if $\lvert z \rvert \geq z_{\alpha/2}$
μ	σ unknown test statistic: $T = \dfrac{(\bar{X} - \mu_0)}{S_X/\sqrt{n}}$	$H_0 : \mu = \mu_0$ $H_1 : \mu \neq \mu_0$	Reject H_0 if $\lvert t \rvert \geq t(n-1, \alpha/2)$
σ^2	$\mu = \mu_0$ known test statistic: $X^2 = \displaystyle\sum_{i=1}^{n} \left(\dfrac{X_i - \mu_0}{\sigma_0} \right)^2$	$H_0 : \sigma^2 = \sigma_0^2$ $H_1 : \sigma^2 \neq \sigma_0^2$	Reject H_0 if $\chi^2 \leq \chi^2(n, 1-\alpha/2)$ or $\chi^2 \geq \chi^2(n, \alpha/2)$
σ^2	μ unknown test statistic: $X^2 = \displaystyle\sum_{i=1}^{n} \left(\dfrac{X_i - \bar{X}}{\sigma_0} \right)^2$ $\qquad = \dfrac{(n-1)S_X^2}{\sigma_0^2}$	$H_0 : \sigma^2 = \sigma_0^2$ $H_1 : \sigma^2 \neq \sigma_0^2$	Reject H_0 if $\chi^2 \leq \chi^2(n-1, 1-\alpha/2)$ or $\chi^2 \geq \chi^2(n-1, \alpha/2)$

Two-sided tests for normal parameters

Parameter	Conditions	Test	Decision Rule
$d = \mu_X - \mu_Y$	$\sigma_X^2 = \sigma_{0,X}^2$ and $\sigma_Y^2 = \sigma_{0,Y}^2$ known test statistic: $Z = \dfrac{(\bar{X}_m - \bar{Y}_n) - d_0}{\sqrt{\frac{\sigma_{0,X}^2}{m} + \frac{\sigma_{0,Y}^2}{n}}}$	$H_0 : d = d_0$ $H_1 : d > d_0$	Reject H_0 if $z \geq z_\alpha$
		$H_0 : d = d_0$ $H_1 : d < d_0$	Reject H_0 if $z \leq -z_\alpha$
$d = \mu_X - \mu_Y$	σ_X^2, σ_Y^2 unknown, $\sigma_X^2 = \sigma_Y^2$ test statistic: $T = \dfrac{(\bar{X}_m - \bar{Y}_n) - d_0}{S_p \sqrt{\frac{1}{m} + \frac{1}{n}}}$, $S_p^2 = \dfrac{(m-1)S_X^2 + (n-1)S_Y^2}{m+n-2}$ (pooled variance)	$H_0 : d = d_0$ $H_1 : d > d_0$	Reject H_0 if $t \geq t(m + n - 2, \alpha)$
		$H_0 : d = d_0$ $H_1 : d < d_0$	Reject H_0 if $t \leq -t(m + n - 2, \alpha)$
$d = \mu_X - \mu_Y$	σ_X^2, σ_Y^2 unknown, $\sigma_X^2 \neq \sigma_Y^2$ test statistic: $T = \dfrac{(\bar{X}_m - \bar{Y}_n) - d_0}{\sqrt{\frac{S_X^2}{m} + \frac{S_Y^2}{n}}}$	$H_0 : d = d_0$ $H_1 : d > d_0$	Reject H_0 if $t \geq t(v, \alpha)$, $v = \dfrac{\left(\frac{1}{m}r + \frac{1}{n}\right)^2}{\frac{1}{m^2(m-1)}r^2 + \frac{1}{n^2(n-1)}}$, $r = \dfrac{\frac{s_X^2}{m}}{\frac{s_Y^2}{n}}$
		$H_0 : d = d_0$ $H_1 : d < d_0$	Reject H_0 if $t \leq -t(v, \alpha)$, $v = \dfrac{\left(\frac{1}{m}r + \frac{1}{n}\right)^2}{\frac{1}{m^2(m-1)}r^2 + \frac{1}{n^2(n-1)}}$, $r = \dfrac{\frac{s_X^2}{m}}{\frac{s_Y^2}{n}}$
$r = \dfrac{\sigma_X^2}{\sigma_Y^2}$	μ_X, μ_Y unknown test statistic: $F = \dfrac{S_X^2}{S_Y^2} \cdot \dfrac{1}{r_0}$	$H_0 : r = r_0$ $H_1 : r > r_0$	Reject H_0 if $f \geq F(m - 1, n - 1, \alpha)$
		$H_0 : r = r_0$ $H_1 : r < r_0$	Reject H_0 if $f \leq F(m - 1, n - 1, 1 - \alpha)$
μ_D (paired samples)	σ_D^2 unknown test statistic: $T = \dfrac{\bar{D} - d_0}{S_D / \sqrt{m}}$	$H_0 : \mu_D = d_0$ $H_1 : \mu_D > d_0$	Reject H_0 if $t \geq t(m - 1, \alpha)$
		$H_0 : \mu_D = d_0$ $H_1 : \mu_D < d_0$	Reject H_0 if $t \leq -t(m - 1, \alpha)$

One-sided tests for the difference of two normal parameters

Parameter	Conditions	Test	Decision Rule
$d = \mu_X - \mu_Y$	$\sigma_X^2 = \sigma_{0,X}^2$ and $\sigma_Y^2 = \sigma_{0,Y}^2$ known test statistic: $Z = \dfrac{(\bar{X}_m - \bar{Y}_n) - d_0}{\sqrt{\frac{\sigma_{0,X}^2}{m} + \frac{\sigma_{0,Y}^2}{n}}}$	$H_0 : d = d_0$ $H_1 : d \neq d_0$	Reject H_0 if $|z| \geq z_{\alpha/2}$
$d = \mu_X - \mu_Y$	σ_X^2, σ_Y^2 unknown, $\sigma_X^2 = \sigma_Y^2$ test statistic: $T = \dfrac{(\bar{X}_m - \bar{Y}_n) - d_0}{S_p \sqrt{\frac{1}{m} + \frac{1}{n}}}$, $S_p^2 = \dfrac{(m-1)S_X^2 + (n-1)S_Y^2}{m+n-2}$ (pooled variance)	$H_0 : d = d_0$ $H_1 : d \neq d_0$	Reject H_0 if $|t| \geq t(m + n - 2, \alpha/2)$
$d = \mu_X - \mu_Y$	σ_X^2, σ_Y^2 unknown, $\sigma_X^2 \neq \sigma_Y^2$ test statistic: $T = \dfrac{(\bar{X}_m - \bar{Y}_n) - d_0}{\sqrt{\frac{S_X^2}{m} + \frac{S_Y^2}{n}}}$	$H_0 : d = d_0$ $H_1 : d \neq d_0$	Reject H_0 if $|t| \geq t(v, \alpha/2)$, $v = \left\lfloor \dfrac{\left(\frac{1}{m}r + \frac{1}{n}\right)^2}{\frac{1}{m^2(m-1)}r^2 + \frac{1}{n^2(n-1)}} \right\rfloor$, $r = \dfrac{s_X^2}{s_Y^2}$
$r = \dfrac{\sigma_X^2}{\sigma_Y^2}$	μ_X, μ_Y unknown test statistic: $F = \dfrac{s_X^2}{s_Y^2} \cdot \dfrac{1}{r_0}$	$H_0 : r = r_0$ $H_1 : r \neq r_0$	Reject H_0 if $f \geq F(m - 1, n - 1, \alpha/2)$ or $f \leq F(m - 1, n - 1, 1 - \alpha/2)$
μ_D (paired samples)	σ_D^2 unknown test statistic: $T = \dfrac{\bar{D} - d_0}{S_D/\sqrt{m}}$	$H_0 : \mu_D = d_0$ $H_1 : \mu_D \neq d_0$	Reject H_0 if $t \geq t(m - 1, \alpha/2)$ or $t \leq -t(m - 1, \alpha/2)$

Two-sided tests for the difference of two normal parameters

Parameter	Conditions	Test	Decision Rule
p	approximate test test statistic: $Z = \dfrac{\bar{X} - p_0}{\sqrt{\frac{p_0(1-p_0)}{n}}}$	$H_0 : p = p_0$ $H_1 : p \neq p_0$	Reject H_0 if $\lvert z \rvert \geq z_{\alpha/2}$
		$H_0 : p = p_0$ $H_1 : p > p_0$	Reject H_0 if $z \geq z_\alpha$
		$H_0 : p = p_0$ $H_1 : p < p_0$	Reject H_0 if $z \leq -z_\alpha$
$d = p_X - p_Y$	approximate test test statistic: $Z = \dfrac{(\bar{X}_m - \bar{Y}_n) - d_0}{\sqrt{\frac{\bar{X}_m(1-\bar{X}_m)}{m} + \frac{\bar{Y}_n(1-\bar{Y}_n)}{n}}}$,	$H_0 : d = d_0$ $H_1 : d \neq d_0$	Reject H_0 if $\lvert z \rvert \geq z_{\alpha/2}$
		$H_0 : d = d_0$ $H_1 : d > d_0$	Reject H_0 if $z \geq z_\alpha$
		$H_0 : d = d_0$ $H_1 : d < d_0$	Reject H_0 if $z \leq -z_\alpha$
$d = p_X - p_Y$	approximate test test statistic: $Z = \dfrac{(\bar{X}_m - \bar{Y}_n)}{\sqrt{\bar{P}_0(1-\bar{P}_0)}\sqrt{\frac{1}{m} + \frac{1}{n}}}$, $\bar{P}_0 = \frac{m\bar{X}_m + n\bar{Y}_n}{m+n}$ (pooled proportion)	$H_0 : d = 0$ $H_1 : d \neq 0$	Reject H_0 if $\lvert z \rvert \geq z_{\alpha/2}$
		$H_0 : d = 0$ $H_1 : d > 0$	Reject H_0 if $z \geq z_\alpha$
		$H_0 : d = 0$ $H_1 : d < 0$	Reject H_0 if $z \leq -z_\alpha$

Tests for binomial proportions

CHAPTER 6

Linear Regression

6.1   INTRODUCTION AND SCATTER PLOTS

In this chapter we discuss one of the most important topics in statistics. It provides us with a way to determine an algebraic relationship between two variables x and the random variable Y which depends on x and ultimately to use the relationship to predict one variable from knowledge of the other. For instance, is there a relationship between a student's GPA in high school and the GPA in college? Is there a connection between a father's intelligence and a daughter's? Can we predict the price of a stock from the S&P 500 index? These are all matters regression can consider.

We have a data set of pairs of points $\{(x_i, y_i), i = 1, 2 \ldots, n\}$. The first step is to create a **scatterplot** of the points. For example, we have the plots in Figures 6.1 and 6.2.

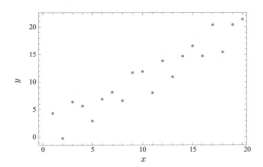

Figure 6.1: Scatterplot of x-y data.

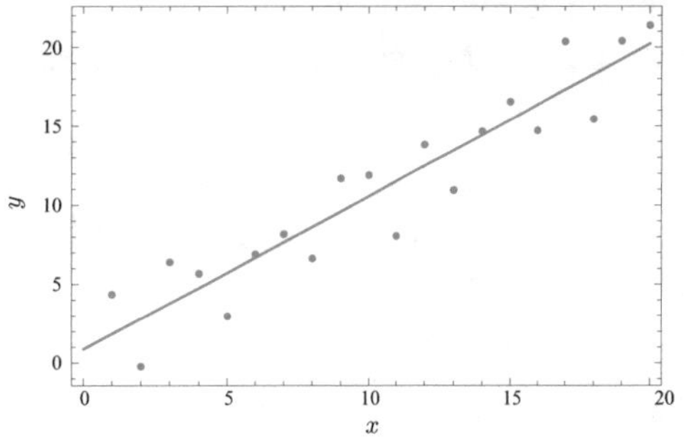

Figure 6.2: Scatterplot of x-y data with line fit to the data.

It certainly looks like there is some kind of relationship between x and y and it looks like it could be linear. By eye we can draw a line that looks like it would be a pretty good fit to the data.

The questions which we will answer in this chapter are as follows.

- How do we measure how good using a line to approximate the data will be?

- How do we find the line which approximates the data as well as possible?

- How do we use the line to make predictions and quantify the errors?

First we turn to the problem of finding the best fit line to the data.

6.2   INTRODUCTION TO REGRESSION

If we have no information about an rv Y, the **best estimate** for Y is EY. The reason for this is the following fact:

$$\min_a E(Y - a)^2 = E(Y - EY)^2 = Var(Y).$$

In other words, EY minimizes the **mean square distance** of the values of Y to any number a. We have seen this to be true earlier but here is a short recap.

First consider the real valued function $f(a) = E(Y - a)^2$. To minimize this function take a derivative and set to zero.

$$f'(a) = -2E(Y - a) = 0 \implies a = EY.$$

Since $f''(a) = 2 > 0$, this $a = EY$ provides the minimum. We conclude that if we have no information about Y's distribution and we have to guess something about Y, then $a = EY$ is the best guess.

Now suppose we know that there is another rv X which is related to Y and we assume that it is a linear relationship. Think of X as the independent variable and Y as the dependent variable, i.e., X is the input and Y is the response. We would really like to precisely describe the linear relationship between X and Y. To do so, we will find constants a, b to minimize the following function giving the mean squared distance of Y to $a + b\,X$:

$$f(a, b) = E(Y - a - bX)^2.$$

To minimize this function (which depends on two variables) take partial derivatives and set to zero:

$$f_a = -2E(Y - a - bX) = 0, \text{ and } f_b = -2E[X(Y - a - bX)] = 0.$$

Solving these simultaneous two equations for a, b we get

$$b = \frac{E(XY) - EXEY}{E(X^2) - (EX)^2} = \frac{E(XY) - EXEY}{Var(X)} \text{ and } a = EY - b\,EX. \tag{6.1}$$

Now we rewrite these solutions using the covariance. Recall that the covariance of X, Y is given by

$$Cov(X, Y) = E(X - EX)(Y - EY) = E(XY) - EXEY$$

and we can rewrite the slope parameter as $b = \frac{Cov(X,Y)}{\sigma_X^2}$. We have

$$\boxed{b = \frac{Cov(X, Y)}{\sigma_X^2} \text{ and } a = EY - bEX.}$$

We can also rewrite the slope b using the **correlation coefficient** $\rho(X, Y) = \frac{Cov(X,Y)}{\sigma_X \sigma_Y}$. We have

$$b = \frac{Cov(X,Y)}{\sigma_X^2} = \frac{Cov(X,Y)}{\sigma_X \sigma_Y} \frac{\sigma_Y}{\sigma_X} = \rho(X,Y) \frac{\sigma_Y}{\sigma_X}.$$

This gives us the result that $b = \rho \frac{\sigma_Y}{\sigma_X}$. We summarize these results.

Proposition 6.1 The minimum of $f(a,b) = E(Y - a - bX)^2$ over all possible constants a, b is provided by $b = \rho(X,Y) \frac{\sigma_Y}{\sigma_X}, a = EY - bEX$. The minimum value of f is then given by

$$f(a,b) = E(Y - a - bX)^2 = (1 - \rho^2)\sigma_Y^2.$$

Proof. We have already shown the first part and all we have to do is find the value of the function f at the point that provides the minimum. We first plug in the value of $a = EY - b\,EX$ and then rearrange:

$$\begin{aligned}
E(Y - a - bX)^2 &= E[(Y - EY) + b(EX - X)]^2 \\
&= E\left[(Y - EY)^2 + 2b(EX - X)(Y - EY) + b^2(EX - X)^2\right] \\
&= Var(Y) + b^2 Var(X) - 2bCov(X,Y) \\
&= \sigma_Y^2 + \rho^2 \frac{\sigma_Y^2}{\sigma_X^2}\sigma_X^2 - 2\rho \frac{\sigma_Y}{\sigma_X}\rho\,\sigma_Y\sigma_X \quad \text{using } b = \rho\frac{\sigma_Y}{\sigma_X}, \\
&= \sigma_Y^2 + \rho^2\sigma_Y^2 - 2\rho^2\sigma_Y^2 = (1 - \rho^2)\sigma_Y^2.
\end{aligned}$$

$\square$

Remark 6.2 The **regression line**, or **least squares fit line**, we have derived is written as

$$Y = a + bX = EY - bEX + bX = EY + b(X - EX) \implies$$

$$\boxed{Y - EY = \rho\frac{\sigma_Y}{\sigma_X}(X - EX).}$$

This shows that the regression line always passes through the point (EX, EY) and has slope $\rho\frac{\sigma_Y}{\sigma_X}$. Consequently, a one standard deviation increase in X from the mean results in a $\rho\,\sigma_Y$ unit increase from the mean (or decrease if $\rho < 0$) in Y.

Example 6.3 Suppose we know that for a rv Y, $\sigma_Y^2 = 10$ and the correlation coefficient between X, Y is $\rho = 0.5$. If we ignore X and try to guess Y, the best estimate is EY which will give us an error of $E(Y - EY)^2 = Var(Y) = 10$. If we use the information on the correlation between X, Y we have instead

$$E[Y - (a + bX)]^2 = (1 - \rho^2)\sigma_Y^2 = (1 - .25)10 = 7.5$$

and the error is cut by 25%.

What we have shown is that the best approximation to the rv Y by a linear function $a + bX$ of a rv X is given by the rv $W = a + bX$ with the constants a, b given by $b = \rho \, \sigma_Y / \sigma_X$ and $a = EY - bEX$. It is not true that $Y = a + bX$ but that $W = a + bX$ is a rv with mean square distance $E(Y - W)^2 = (1 - \rho^2)\sigma_Y^2$, and this is the smallest possible such distance for any possible constants a, b.

Remark 6.4   Since $E(Y - a - bX)^2 = (1 - \rho^2)\sigma_Y^2 \geq 0$, it must be true that $|\rho| \leq 1$. Also, if $|\rho| = 1$, then the only possibility is that $Y = a + bX$ so that Y is exactly a linear function of X.

Now we see that ρ is a quantitative measure of how well a line approximates Y. The closer ρ is to ± 1, the better the approximation. When $|\rho| = 1$ we say that Y is **perfectly linearly correlated** with X. When $\rho = 0$ the error of approximation by a linear function is the largest possible error. When $\rho = 0$ we say that Y **is uncorrelated with** X.

As a general rule, if $|\rho| \leq 0.5$ the **correlation is weak** and when $|\rho| \geq 0.8$ we say the linear **correlation is strong**.

There is another interpretation of ρ^2 through the equation $E(Y - a - bX)^2 = (1 - \rho^2)\sigma_Y^2 \geq 0$. Rearranging this we have

$$\rho^2 = 1 - \frac{E(Y - a - bX)^2}{\sigma_Y^2} = \frac{\sigma_Y^2 - E(Y - a - bX)^2}{\sigma_Y^2}.$$

It is common to refer to ρ^2 as the **coefficient of determination**. It has the interpretation that it represents the proportion of the total variation in the Y-values, explained by the variation in the Y-values due to the linear relationship itself. For instance, if $\sigma_Y^2 = 1500$ and $E(Y - a - bX)^2 = 1100$, then, by Proposition 6.1, $\rho^2 = .2667$ so that 26.67% of the variation in the Y-values is explained by the linear relationship, while about 74% is unexplained. Another way to interpret ρ^2 is that it represents the total proportion of the variation in the Y-values which is reduced by taking into account the predictor value X.

6.2.1   THE LINEAR MODEL WITH OBSERVED X

So far we have assumed that (X, Y) are both random variables with some joint distribution. In many applications X is not a random variable but a fixed observable variable and Y is a random variable whose mean depends in a linear way on the observable $X = x$. The linear model assumes

$$Y = a + bx + \varepsilon, \quad \text{where } \varepsilon \sim N(0, \sigma).$$

The random variable ε represents the random noise at the observation x. (See Figure 6.3.) The mean of Y will change in a linear way with x and for each x the Y values will distribute according to a normal distribution with standard deviation σ. We have with this model

$$EY = E(a + bx + \varepsilon) = a + bx, \text{ and } SD(Y) = \sqrt{E(Y - a - bx)^2} = \sqrt{E(\varepsilon^2)} = \sigma.$$

If we knew σ, then we are saying that for each observed value of x, we have $Y \sim N(a + bx, \sigma)$. Thus, for each fixed observed value x, the mean of Y is $a + bx$ and the Y-values are normally distributed with SD given by σ.

Now suppose the data pairs $(x_i, y_i), i = 1, 2, \ldots, n$, are observations from (x_i, Y_i), where $Y_1, Y_2, \ldots, Y_n$ is a random sample from $Y = y_i = a + b x_i + \varepsilon_i$, with independent errors $\varepsilon_i \sim N(0, \sigma)$.

Example 6.5   Suppose the relationship between interest rates and the value of real estate is given by a simple linear regression model with the regression line $Y = 137.5 - 12.5x + \varepsilon$, where x is the interest rate and Y is the value of the real estate. We suppose the noise term $\varepsilon \sim N(0, 10)$. Then for any fixed interest rate x_0, the real estate will have the distribution $N(-12.5x_0 + 137.5, 10)$. For instance, if $x_0 = 8$, $EY = 37.5$, and then

$$P(Y > 45 | x_0 = 8) = \text{normalcdf}(45, \infty, -12.5(8) + 137.5, 10) = 0.2266.$$

The mean value of real estate when interest rates are 8% is 37.5 and there is about a 22% chance the real estate value will be above 45.

Notice that because the slope of the regression line is negative, higher interest rates will give lower values of real estate. Suppose $Y_1 = -12.5x_1 + 137.5 + \varepsilon_1$ is an observation when $x_1 = 10$ and $Y_2 = -12.5x_2 + 137.5 + \varepsilon_2$ is an observation when $x_2 = 9$. By properties of the normal distribution $Y_1 - Y_2 \sim N(-12.5(x_1 - x_2), 14.14)$. What are the chances real estate values will be higher with rates at 10% rather than at 9%? That is, what is $P(Y_1 > Y_2)$? Here's the answer:

$$P(Y_1 - Y_2 > 0) = \text{normalcdf}(0, \infty, -12.5, 14.14) = 0.1883.$$

There is an 18.83% chance that values will be higher at 10% interest than at 9%. Next, suppose the value of real estate at 9% is actually 35. What percentile is this? That's easy because we are assuming $Y \sim N(25, 10)$ so that $P(Y < 35 | x = 9) = \text{normalcdf}(-\infty, 35, 25, 10) = 0.841$ and so 35 is the 84th percentile of real estate values when interest rates are 9%. This means that 84% of real estate values are below 35 when interest rates are 9%.

The term **regression** implies a return to a less developed state. The question naturally arises as to why this term is applied to linear regression analysis. Informally, Galton, the first developer of regression analysis, noticed that tall fathers had tall sons, but not quite so tall as the father. He termed this as **regression to the mean**, implying that the height of the sons is regressing more toward what the mean height should be for men of a certain age. It was also noticed that students who did very well on an exam would not do quite so well on a retake of the exam. They were regressing to the mean.

The mathematical explanation for this is straightforward from the model $Y = a + bx + \varepsilon$, $\varepsilon \sim N(0, \sigma)$. Suppose we fix the input x and take a measurement Y. This measurement itself follows a normal distribution with mean $a + bx$ and SD σ. Suppose the measurement is $Y = y_1$ and this measurement is 1.5 standard units above average. That would put it at the 93.32 percentile. If that's a test score it's a good score. On another measurement of $Y = y_2$ for the same x, 93.32% of the observations are below y_1. This means there is a really good chance the second observation will be below y_1. There is some chance $y_2 > y_1$, ($\approx 7\%$) but it's unlikely compared to the chance $y_2 < y_1$. That is what regression to the mean refers to. The **regression fallacy** is attributing the change in scores to something important rather than just to the chance variability around the mean.

6.2.2   ESTIMATING THE SLOPE AND INTERCEPT FROM DATA

Now suppose we have observed data points $(x_i, y_i), i = 1, 2, \ldots, n$. If we take the least squares regression line

$$Y - EY = \rho \frac{\sigma_Y}{\sigma_X}(X - EX)$$

and replace each statistical quantity with its estimate using data, we get the best fit line for the data as

$$y - \bar{y} = r \frac{s_Y}{s_X}(x - \bar{x}), \qquad \bar{x} = \frac{1}{n}\sum_{i=1}^{n} x_i, \qquad \bar{y} = \frac{1}{n}\sum_{i=1}^{n} y_i$$

and the **sample correlation coefficient**

$$r = \frac{\frac{1}{n-1}\sum_{i=1}^{n}(x_i - \bar{x})(y_i - \bar{y})}{s_X s_Y} = \frac{1}{n-1}\sum_{i=1}^{n} \frac{x_i - \bar{x}}{s_X} \frac{y_i - \bar{y}}{s_Y},$$

where the usual formulas for the sample variances are

$$s_Y^2 = \frac{1}{n-1}\sum_{i=1}^{n}(y_i - \bar{y})^2, \qquad s_X^2 = \frac{1}{n-1}\sum_{i=1}^{n}(x_i - \bar{x})^2.$$

This says that the sample correlation coefficient is calculated by converting each pair (x_i, y_i) to standard units and then (almost) averaging the products of the standard units.

If we wish to write the regression equation in slope intercept form we have

$$y = \hat{a} + \hat{b}x \quad \text{where} \quad \hat{a} = \bar{y} - \hat{b}\,\bar{x}, \quad \hat{b} = r\frac{s_Y}{s_X}.$$

This is the regression equation derived from the probabilistic model $Y = a + bx + \varepsilon$. If we ignore the probability aspects we can derive the same equations as follows. The proof is a calculus exercise.

Proposition 6.6   Given data points $(x_i, y_i), i = 1, 2, \ldots, n$, set $f(a, b) = \sum_{i=1}^{n}(y_i - a - bx_i)^2$.

Then the minimum of f over all $a \in \mathbb{R}, b \in \mathbb{R}$ is achieved at $\hat{b} = r\dfrac{s_Y}{s_X}$, $\hat{a} = \bar{y} - \hat{b}\bar{x}$ and the minimum is $f(\hat{a}, \hat{b}) = (n - 1)(1 - r^2)s_Y^2$.

Example 6.7   Suppose we choose 11 families randomly and we let $x_i =$ height of brother, $y_i =$ height of sister. We have the summary statistics $\bar{x} = 69, \bar{y} = 64, s_X = 2.72, s_Y = 2.569$. The correlation coefficient is given by $r = 0.558$. The equation of the regression line is therefore

$$y - 64 = 0.558\frac{2.569}{2.72}(x - 69) = 0.527(x - 69).$$

If a brother's height is actually $69 + 2.72$, then the sister's mean height will be $64 + 0.558(2.569) = 65.43$. The minimum squared error is $f(27.637, 0.527) = 10(1 - 0.558^2)2.569^2 = 45.448$. Any other choice of a, b will result in a larger error.

Example 6.8   Suppose we know that a student scored in the 82nd percentile of the SAT exam in high school and we know that the correlation between high school SAT scores and first-year college GPA is $\rho = .9$. Assuming that high school SAT scores and GPA scores are normally distributed, what will this student's predicted percentile GPA be?

This problem seems to not provide enough information. Shouldn't we know the means and SDs of the SAT and GPA scores? Actually we do have enough information to solve it. First, rewrite the regression equation as

$$\frac{y - \bar{y}}{s_Y} = r\frac{x - \bar{x}}{s_X},$$

where $x = SAT$ and $y = GPA$. Knowing that the student scored in the 82nd percentile of SAT scores, which is assumed normally distributed, tells us that this student's SAT score in standard units is $\text{invNorm}(.82) = 0.9154 = \dfrac{x - \bar{x}}{s_X}$. Therefore, the student's GPA score in standard units is $\dfrac{y - \bar{y}}{s_Y} = 0.9(0.9154) = 0.8239$. Therefore, the student's predicted percentile for GPA is $\text{normalcdf}(-\infty, .8239) = 0.795$, or 79.5 percentile.

6.2.3   ERRORS OF THE REGRESSION

Next, we need to go further into the use of the regression line to predict y values for given x values and to determine how well the line approximates the data. We will present an ANOVA for linear regression to decompose the variation in the dependent variable.

We use the notation

$$SST = E(Y - EY)^2 = \sigma_Y^2, \quad SSE = E(Y - a - bX)^2, \quad SSR = E(a + bX - EY)^2.$$

SST is the total variation in the Y-values and is decomposed in the next proposition into the variation of the Y-values from the fitted line, (SSE), and the variance of the fitted values to the mean of Y, (SSR).

Proposition 6.9   Let $b^* = \rho\frac{\sigma_Y}{\sigma_X}$ and $a^* = EY - b^*EX$. Then

$$SST \equiv \sigma_Y^2 = \underbrace{E(Y - EY)^2}_{SST} = \underbrace{E(Y - a^* - b^* X)^2}_{SSE} + \underbrace{E(a^* + b^* X - EY)^2}_{SSR} \equiv SSE + SSR.$$

$$(6.2)$$

Proof.

$$\sigma_Y^2 = E(Y - EY)^2 = E(Y - a^* - b^* X + a^* + b^* X - EY)^2$$
$$= E(Y - a^* - b^* X)^2 + 2E(Y - a^* - b^* X)(a^* + b^* X - EY) + E(EY - a^* - b^* X)^2$$
$$= SSE + 0 + SSR.$$

The middle term is zero because

$$E(Y - a^* - b^* X)(a^* + b^* X - EY) = Eb^*(Y - EY)(X - EX) - (b^*)^2 E(X - EX)^2$$

$$= E\rho\frac{\sigma_Y}{\sigma_X}(Y - EY)(X - EX) - \rho^2\frac{\sigma_Y^2}{\sigma_X^2}\sigma_X^2 = \rho\frac{\sigma_Y}{\sigma_X}\rho\sigma_Y\sigma_X - \rho^2\sigma_Y^2 = 0.$$

$\square$

When we have data points (x_i, y_i), SSE is the variation of the residuals $(y_i - a^* - bx_i^*)^2$. SSR is the variation due to the regression $(\bar{y} - a^* - b^*x_i)^2$. Using this decomposition we can get some important consequences.

Proposition 6.10   We have (a) $SSE = (1 - \rho^2)SST$, and,   (b) $\rho^2 = \dfrac{SSR}{SST} = 1 - \dfrac{SSE}{SST}$.

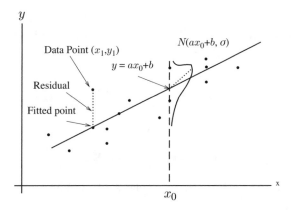

Figure 6.3: Model for linear regression.

Proof. To show (a), using Proposition 6.1

$$SSE = E(Y - a^* - b^* X)^2 = (1 - \rho^2) \sigma_Y^2 = (1 - \rho^2) SST.$$

Using (a),

$$SST = SSE + SSR = SST(1 - \rho^2) + SSR \implies \rho^2 = \frac{SSR}{SST} = \frac{SST - SSE}{SST} = 1 - \frac{SSE}{SST}.$$

$\square$

Now suppose we are given a line $y = a + bx$ and we have data points $\{(x_i, y_i)\}$. We will set $\hat{b} = r \frac{s_Y}{s_X}$ and $\hat{a} = \bar{y} - \hat{b}\,\bar{x}$ as the coefficients of the best fit line when we have the data.

Definition 6.11 The **fitted values** are $\hat{y}_i = \hat{a} + \hat{b}\,x_i$. These are the points on the regression line for the associated x_i. **Residuals** are $\varepsilon_i = y_i - \hat{y}_i$, the difference between the observed data values and the fitted values.

It is always true for the regression line $y = \hat{a} + \hat{b}\,x$ that $\sum \varepsilon_i = 0$ since $\hat{a} = \bar{y} - \hat{b}\,\bar{x}$ (Figure 6.3). Therefore basing errors on the sum of the residuals won't work.

We have the observed quantities y_i, the calculated quantities $\hat{y}_i = \hat{y}_i(x_i)$, and the residuals $\varepsilon_i = y_i - \hat{y}_i$ which is the amount by which the observed value differs from the fitted value. The residuals are labeled ε_i because $\varepsilon_i = y_i - \hat{a} - \hat{b}x_i$.

Another way to think of the ε_i's is as observations from the normal distribution giving the errors at each x_i. We have chosen the line so that

$$\sum_{i=1}^{n} \varepsilon_i^2 = \sum_{i=1}^{n} (y_i - \hat{y}_i)^2 = \sum_{i=1}^{n} \left(y_i - \hat{a} - \hat{b}x_i\right)^2$$

is the minimum possible.

In the preceding definitions of SST, SSE, and SSR when we have data points (x_i, y_i) we replace Y by y_i, X by x_i, and ρ by r. We have:

(a) Error sum of squares, deviation of y_i from $\hat{y}_i$: $SSE = \sum_{i=1}^{n}(y_i - \hat{y}_i)^2 = \sum_{i=1}^{n}\varepsilon_i^2$.

(b) Regression sum of squares, deviation of $\hat{y}_i$ from $\overline{y}$: $SSR = \sum_{i=1}^{n}(\hat{y}_i - \overline{y})^2$. This is the amount of total variation in the y-values that is explained by the linear model.

(c) Total sum of squares, deviation of data values y_i from $\overline{y}$: $SST = \sum_{i=1}^{n}(y_i - \overline{y})^2$. In other words $\frac{SST}{n-1}$ is the sample variance of the y-values.

The algebraic relationships between the quantities SST, SSR, SSE become

$$SST = SSR + SSE \text{ and } r^2 = \frac{SSR}{SST} = 1 - \frac{SSE}{SST} \text{ and } SSE = \left(1 - r^2\right)SST.$$

We will use the notation

$$S_{yy} = \sum_{i=1}^{n}(y_i - \overline{y})^2 = SST \quad S_{xx} = \sum_{i=1}^{n}(x_i - \overline{x})^2 \quad S_{xy} = \sum_{i=1}^{n}(x_i - \overline{x})(y_i - \overline{y})$$

$$\hat{a} = \overline{y} - \hat{b}\,\overline{x}, \quad \hat{b} = r\frac{s_Y}{s_X} = \frac{S_{xy}}{S_{xx}}.$$

The expression for the slope follows from the computation

$$\hat{b} = r\frac{s_Y}{s_X} = \frac{\frac{1}{n-1}\sum_{i=1}^{n}(x_i - \overline{x})(y_i - \overline{y})}{s_X s_Y}\frac{s_Y}{s_X} = \frac{S_{xy}}{S_{xx}}.$$

Remark 6.12   For computation, the following formulas can simplify the work:

$$S_{xx} = \sum x_i^2 - n\overline{x}^2, \quad S_{yy} = \sum y_i^2 - n\overline{y}^2, \quad S_{xy} = \sum x_i y_i - n\overline{x}\,\overline{y}.$$

The Estimate of σ^2

Recall that we assumed $Y = \hat{a} + \hat{b}x + \varepsilon$, where $\varepsilon \sim N(0, \sigma)$. We have for fixed x the mean of the data given x is $E(Y|x) = \hat{a} + \hat{b}x$ and $Var(Y) = \sigma^2$. The variance σ^2 measures the spread of the data around the mean $\hat{a} + \hat{b}x$. The estimate of σ^2 is given by

$$s^2 = \frac{\sum_{i=1}^{n}(y_i - \hat{y}_i)^2}{n-2} = \frac{SSE}{n-2}.$$

The sample value s is called the **standard error of the estimate** and represents the deviation of the y data values from the corresponding fitted values of the regression line. We will see later that $S^2 = \frac{1}{n-2}\sum_{i=1}^{n}(Y_i - \hat{a} - \hat{b}x_i)^2$ is an unbiased estimator of σ^2, and S^2 will be associated with a $\chi^2(n-2)$ random variable. The degrees of freedom will be $n-2$ because of the involvement of two unknown parameters $\hat{a}, \hat{b}$.

The value of s measures how far above or below the data points are from the regression line. Since we are assuming a linear model with noise which is normally distributed, we can say that roughly 68% of the data points will lie within the band created by two parallel lines to the regression line, one above the line and one below. If the width of this band is $2s$, it will contain roughly 95% of all the data points. Any data point lying outside the $2s$ band is considered an **outlier**.

Remark 6.13   We have shown in Proposition 6.6 that given data points $(x_i, y_i), i = 1, 2, \ldots, n$, if we set

$$f(a, b) = \sum_{i=1}^{n}(y_i - a - bx_i)^2,$$

then the minimum of f over all $a \in \mathbb{R}, b \in \mathbb{R}$ is achieved at $\hat{b} = r\frac{s_Y}{s_X}$, $\hat{a} = \bar{y} - \hat{b}\bar{x}$ and the minimum is $f\left(\hat{a}, \hat{b}\right) = (n-1)(1-r^2)s_Y^2$. It is obvious from the definitions that

$$f\left(\hat{a}, \hat{b}\right) = (n-1)(1-r^2)s_Y^2 = \sum_{i=1}^{n}\left(y_i - \hat{a} - \hat{b}x_i\right)^2 = \sum_{i=1}^{n}\varepsilon_i^2.$$

Thus, a simple formula for the SE of the regression is

$$s = \sqrt{\frac{SSE}{n-2}} = \sqrt{\frac{n-1}{n-2}}s_Y\sqrt{1-r^2}.$$

Example 6.14   The table contains data for the pairs (height of father, height of son) as well as the predicted (mean) height of the son for each given height of the father. The resulting difference of the observed height of the son from the prediction, i.e., the residual, is also listed. The residuals should not exhibit a consistent pattern but be both positive and negative. Otherwise, a line may be a bad fit to the data.

X, Father	65	63	67	64	68	62
Y, Son	68	66	68	65	69	66
$\hat{y}$, predicted	66.79	65.84	67.74	66.31	68.22	65.36
ε, residuals	1.21	0.16	0.26	-1.31	0.78	0.64
X, Father	70	66	68	67	69	71
Y, Son	68	65	71	67	68	70
$\hat{y}$, predicted	69.17	67.27	68.22	67.74	68.69	69.65
ε, residuals	-1.17	-2.27	2.72	-0.74	-0.69	0.35

The regression line is $y = 35.82480 + 0.476377\, x$ and the sample correlation coefficient is $r = 0.70265$. Also, $s_X = 2.7743, \overline{x} = 66.67, s_Y = 1.8809, \overline{y} = 67.583$. The coefficient of determination is $r^2 = 0.4937$, which means 49% of the variation in the y-values is explained by the regression. The standard error of the estimate of σ is $s = 1.40366$ which may be calculated using $s = 1.8809 \times \sqrt{11/10} \times \sqrt{1 - 0.4937}$.

6.3   THE DISTRIBUTIONS OF $\hat{a}$ AND $\hat{b}$

If we look at the model $Y = a + bx + \varepsilon, \varepsilon \sim N(0, \sigma)$, from a probabilistic point of view we derived that $\hat{b} = r(x, Y)\frac{s_Y}{s_x}$ and $\hat{a} = \overline{Y} - \hat{\beta}\overline{x}$. Our point of view is that $\hat{a}$ and $\hat{b}$ are estimated from a random sample of $Y_1, \ldots, Y_n$ associated with the fixed deterministic values $x_1, \ldots, x_n$. Thus, from this point of view, $\hat{a}$ **and** $\hat{b}$ **are random variables** and we need to know their distributions to estimate various errors. The following theorem summarizes the results.

Theorem 6.15   The slope $\hat{b}(x_1, \ldots, x_n, Y_1, \ldots, Y_n)$ and intercept $\hat{a}(x_1, \ldots, x_n, Y_1, \ldots, Y_n)$ both have a normal distribution. In fact,

$$\hat{a} \sim N\left(a, \sigma\sqrt{\frac{\sum_{i=1}^{n} x_i^2}{n\,S_{xx}}}\right), \quad \text{and} \quad \hat{b} \sim N\left(b, \frac{\sigma}{\sqrt{S_{xx}}}\right).$$

Proof. We will only show the result for $\hat{b}$. We have from the fact that $\sum_{i=1}^{n}(x_i - \overline{x}) = n\overline{x} - n\overline{x} = 0$,

$$\hat{b} = \frac{S_{xY}}{S_{xx}} = \frac{\sum_{i=1}^{n}(x_i - \overline{x})(Y_i - \overline{Y})}{S_{xx}} = \sum_{i=1}^{n} \frac{x_i - \overline{x}}{S_{xx}} Y_i.$$

Remember that Y_i here is random while x_i is deterministic. This computation shows that $\hat{b}$ is a linear combination of independent normally distributed random variables and therefore $\hat{b}$ also has a normal distribution. Now to calculate the mean and SD of $\hat{b}$ we compute

$$E[\hat{b}] = \sum_{i=1}^{n} \frac{x_i - \overline{x}}{S_{xx}} E[Y_i] = \sum_{i=1}^{n} \frac{x_i - \overline{x}}{S_{xx}}(a + bx_i) = b,$$

where we use the facts $\sum \frac{x_i - \overline{x}}{S_{xx}} = 0$ and $\sum(x_i - \overline{x})x_i = \sum(x_i - \overline{x})^2 = S_{xx}$. To find the SD of $\hat{b}$, we have from the independence of the random variables Y_i,

$$Var[\hat{b}] = \sum_{i=1}^{n} \left(\frac{x_i - \overline{x}}{S_{xx}}\right)^2 Var(Y_i) = \sigma^2 \frac{1}{S_{xx}^2} S_{xx} = \frac{\sigma^2}{S_{xx}}.$$

For $\hat{a} = \overline{Y} - \hat{b}\,\overline{x}$, we have $E(\hat{a}) = E\overline{Y} - b\,\overline{x} = a + b\,\overline{x} - b\,\overline{x} = a$. Also, assuming $\overline{Y}$ and $\hat{b}$ are independent rvs (which we skip showing), we see that $Var(\hat{a}) = Var(\overline{Y}) + \overline{x}^2 Var(\hat{b}) = \frac{\sigma^2}{n} + \overline{x}^2 \frac{\sigma^2}{S_{xx}}$. A little algebra gives the result for the variance. Finally, $\overline{Y}$ and $\hat{b}$ are normal and independent so that $\hat{a}$ is also normal. $\qquad\qquad\square$

The problem with this result is that the distributions depend on the unknown parameter σ. As we usually do in statistics, we replace σ^2 with its estimate $s^2 = \frac{1}{n-2}\sum(y_i - \hat{y}_i)^2$. The random variable analogue of this is

$$S^2 = \frac{1}{n-2} \sum_{i=1}^{n} \left(Y_i - \hat{a} - \hat{b}\,x_i\right)^2 = \frac{1}{n-2}\sum_{i=1}^{n} \varepsilon_i^2 = \frac{SSE}{n-2}.$$

Since $\varepsilon_i \sim N(0, \sigma), i = 1, 2, \ldots, n$, are independent and normally distributed, the sum of squares of normals has a χ^2 distribution. The numerator of S^2 seems to be $\chi^2(n)$. However, there are two parameters $\hat{a}$ and $\hat{b}$ in this expression so the degrees of freedom actually turns out to be $n - 2$, not n. That is,

$$\boxed{\frac{n-2}{\sigma^2} S^2 = \frac{SSE}{\sigma^2} \sim \chi^2(n-2).}$$

This means that if we replace σ by s in the distributions of $\hat{a}$ and $\hat{b}$, we have the following.

Theorem 6.16

$$\boxed{\frac{\hat{a} - a}{S\sqrt{\dfrac{\sum_{i=1}^{n} x_i^2}{n S_{xx}}}} \sim t(n-2) \quad \text{and} \quad \frac{\hat{b} - b}{S/\sqrt{S_{xx}}} \sim t(n-2).}$$

Proof. We have the standardized random variables

$$\frac{\hat{a} - a}{SD(\hat{a})} = \frac{\hat{a} - a}{\sigma\sqrt{\dfrac{\sum_{i=1}^{n} x_i^2}{n S_{xx}}}} \sim N(0,1) \text{ and } \frac{\hat{b} - b}{SD(\hat{b})} = \frac{\hat{b} - b}{\sigma/\sqrt{S_{xx}}} \sim N(0,1).$$

Therefore, since $S = \sigma\sqrt{\chi^2(n-2)/(n-2)}$

$$\frac{\hat{a} - a}{S\sqrt{\dfrac{\sum_{i=1}^{n} x_i^2}{n S_{xx}}}} = \frac{\dfrac{\hat{a} - a}{\sigma\sqrt{\dfrac{\sum_{i=1}^{n} x_i^2}{n S_{xx}}}}}{S/\sigma} = \frac{N(0,1)}{\sqrt{\dfrac{\chi^2(n-2)}{n-2}}} = t(n-2).$$

The computation for $\hat{b}$ is similar. □

As usual, when we replace σ by the sample standard deviation, the normal distribution gets changed to a t-distribution.

6.4   CONFIDENCE INTERVALS FOR SLOPE AND INTERCEPT AND HYPOTHESIS TESTS

Now that we have the Standard Errors for the slope and intercept of a linear regression, we may construct confidence intervals for these parameters and perform hypothesis tests. With an abuse of notation, we will use $\hat{a} = \hat{a}((x_1, y_1), \ldots, (x_n, y_n)) = \overline{y} - \hat{b}\overline{x}$ and $\hat{b} = \hat{b}((x_1, y_1), \ldots, (x_n, y_n)) = r\frac{s_Y}{s_X}$ to denote the intercept and slope, respectively, when we have data points $(x_i, y_i), i = 1, 2, \ldots, n$. That is, they are not random variables in discussing CIs.

Confidence Intervals: The $100(1 - \alpha)\%$ confidence interval for the intercept a is

$$\hat{a} \pm t(n-2, \alpha/2)SE(\hat{a}) = \hat{a} \pm t(n-2, \alpha/2)s\sqrt{\frac{\sum x_i^2}{n S_{xx}}},$$

and the slope b is

$$\hat{b} \pm t(n-2, \alpha/2)SE(\hat{b}) = \hat{b} \pm t(n-2, \alpha/2)\frac{s}{\sqrt{S_{xx}}}.$$

Example 6.17   Consider the data from Example 6.14 concerning heights of fathers and sons. We calculate a 95% CI for $\hat{a}$ and $\hat{b}$. We have $\hat{a} = 35.8248, \hat{b} = 0.4764$. Using Theorem 6.16

we calculate using $SE(\hat{a}) = s\sqrt{\frac{\sum x_i^2}{ns_{xx}}}$, that $s = \sqrt{SSE/(n-2)} = \sqrt{19.702/10} = 1.4036$. Since $\sum x_i^2 = 53418$, $\bar{x} = 66.66$, we have $S_{xx} = 53418 - 12 \times 66.66^2 = 84.667$ and then $SE(\hat{b}) = 1.4036/\sqrt{84.667} = 0.1525$. For 95% confidence, we have $t(10, 0.05) = invT(0.975, 10) = 2.228$ and then the CI for the slope is $0.4764 \pm 2.228 \times 0.1525$, i.e., $(0.136, 0.816)$. The 95% CI for the intercept is $35.8248 \pm 2.228 \times 1.4036\sqrt{\frac{53418}{12 \times 84.67}} = 35.8248 \pm 22.675$.

Hypothesis Tests for Slope and Intercept

Once we know the distributions of the slope and intercept it is straightforward to do hypothesis tests on them. If we specify an intercept a_0 or a slope b_0 the test of the hypothesis that the parameter takes on the specified value is summarized.

- Test for a : $H_0 : a = a_0$, a_0 specified.

  Test Statistic: $t_{a_0} = \dfrac{\hat{a} - a_0}{SE(\hat{a})}$. This is the observed value of the intercept converted to standard units.

$$
\begin{array}{lll}
H_1 : a < a_0 & \text{then, reject null if} & t_{a_0} \leq -t(n-2, \alpha) \\
H_1 : a > a_0 & \text{then, reject null if} & t_{a_0} \geq t(n-2, \alpha) \\
H_1 : a \neq a_0 & \text{then, reject null if} & |t_{a_0}| \geq t_{n-2, \alpha/2}.
\end{array}
$$

- Test for b : $H_0 : b = b_0$, b_0 specified.

  Test Statistic: $t_{b_0} = \dfrac{\hat{b} - b_0}{SE(\hat{b})}$. This is the observed value of the slope converted to standard units.

$$
\begin{array}{lll}
H_1 : b < b_0 & \text{then, reject null if} & t_{b_0} \leq -t(n-2, \alpha) \\
H_1 : b > b_0 & \text{then, reject null if} & t_{b_0} \geq t(n-2, \alpha) \\
H_1 : b \neq b_0 & \text{then, reject null if} & |t_{b_0}| \geq t(n-2, \alpha/2).
\end{array}
$$

- We can use the test for the slope to determine if there is a linear relationship between the x and Y variables. This is based on the fact that $b = \rho\, \sigma_Y/\sigma_X$ so that $\rho = 0$ if and only if $b = 0$. Therefore to test for a linear relationship the null is $H_0 : b = 0$. The alternative hypothesis is that there is no linear relationship, $H_1 : b \neq 0$. The test statistic is $t_0 = \frac{\hat{b}-0}{SE(\hat{b})}$. The null is rejected at level α if $|t_0| > t(n-2, \alpha/2)$.

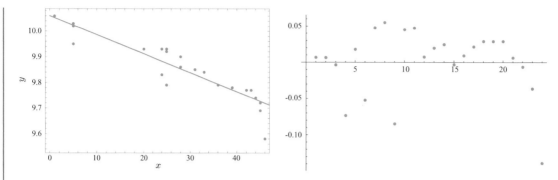

Figure 6.4: World record 100 meter times, 1964–2009.

Figure 6.5: Residuals.

In all cases we may calculate the p-values. For instance, if $H_1 : b < b_0$ in the test for the slope $H_0 : b = b_0$, the p-value is $P(t_{b_0} \leq -t(n-2))$.

Example 6.18   Figure 6.4 (Figure 6.5 is a plot of the residuals) is a plot of the world record 100 meter times (for men) measured from 1963 until 2009. There are 24 data points with the last data point $(46, 9.58)$ corresponding to $(2009, 9.58)$. This point beat the previous world record $(2008, 9.69)$ by 8 seconds!

The regression line is given by $y = 10.0608 - 0.00743113x$ with a correlation coefficient of $r = -0.91455$. It is easy to calculate $SSR = 0.256318, SSE = 0.050139, SST = 0.030645$, and $SE(\hat{a}) = 0.0227721, SE(\hat{b}) = 0.0000700655$. The standard error of the estimate of the regression is $s = \sqrt{SSE/22} = 0.047735$. Also, $S_{xx} = 4641.625$. The 95% CI for the slope is $-0.0074313 \pm t(22, 0.025)\frac{0.047735}{\sqrt{4641.625}}$ which gives $(-0.00888, -0.005978)$. If we test the hypothesis $H_0 : b = 0$ the test statistic is $t = \frac{-0.00743113}{.047735/\sqrt{4641.625}} = -10.606$. Since $t(22, 0.025) = 2.818$ we will reject the null if $\alpha = 0.01$. Observe also that if we project the linear model into the future, in the year 3316, the world record time would be zero.

ANOVA for Linear Regression

We have encountered this topic when we considered hypothesis testing in Chapter 5. Analysis of variance (ANOVA) in regression is a similar method to decompose the variation in the y values into the components determined by the source of the variation. For instance, the simple formula $SST = SSR + SSE$ we derived earlier is an example of the decomposition. As in Chapter 5, we will exhibit the ANOVA in tabular form:

Source of Variation	Degrees of Freedom	Sum of Squares	Mean Square	F-statistic
Regression	1	$SSR = \sum_{i=1}^{n}(\hat{y}_i - \bar{y})^2$	$MSR = \frac{SSR}{1}$	$\frac{MSR}{MSE} = f(1, n-2)$
Residuals	$n-2$	$SSE = \sum_{i=1}^{n}(y_i - \hat{y}_i)^2$	$MSE = \frac{SSE}{n-2}$	
Total	$n-1$	$SST = \sum_{i=1}^{n}(y_i - \bar{y})^2$		

The degrees of freedom for SST is $n-1$ because $SST = \sum_{i=1}^{n}(y_i - \bar{y})^2$ is the sum of n squares subject to one constraint $\sum(y_i - \bar{y}) = 0$ which eliminates one degree of freedom.

By definition, if we take a sum of squares and divide by its degree of freedom we call that the Mean Square. Therefore,

$$MST = SST/(n-1), \quad MSR = SSR/1, \quad MSE = SSE/(n-2).$$

Since $s^2 = SSE/(n-2)$ is the estimate of σ^2, we have $MSE = s^2$.

Next consider the ratio

$$\frac{MSR}{MSE} = \frac{SSR}{\frac{SSE}{n-2}} = \frac{\hat{b}^2 S_{xx}}{s^2} = \left(\frac{\hat{b}}{s/\sqrt{S_{xx}}}\right)^2 = \left(\frac{\hat{b}}{SE(\hat{b})}\right)^2.$$

Notice that $\frac{\hat{b}}{SE(\hat{b})} = t(n-2)$, i.e., it has a t-distribution with $n-2$ degrees of freedom. This is the test statistic for the hypothesis $H_0 : b = 0$. The square of a t-distribution with k degrees of freedom is an $F(1, k)$ distributed random variable. Therefore, the last column says that MSR/MSE has an F-distribution with degrees of freedom $(1, n-2)$. This gives us the value of the test statistic squared in the hypothesis test $H_0 : b = 0$ against $H_1 : b \neq 0$ and we may **reject** H_0 if $F > f(1, n-2, \alpha)$ at level α.

Example 6.19   A delivery company records the following delivery times depending on miles driven.

Distance	2	2	2	5	5	5	10	10	10	15	15	15
Time	10.2	14.6	18.2	20.1	22.4	30.6	30.8	35.4	50.6	60.1	68.4	72.1

The regression line becomes $y = 4.54677 + 3.94728\,x$. The statistics for the regression line are summarized in the table

	Estimate	Standard Error	t-Statistic	p-Value
$\hat{a}$	4.54677	3.8823	1.17115	0.268687
$\hat{b}$	3.94728	0.412684	9.5649	2.386×10^{-6}

For example, $SE(\hat{b}) = 0.412684 = \frac{s}{\sqrt{S_{xx}}}$, since $s = \sqrt{11/10} \times 21.4932 \times \sqrt{1 - .949455^2} = 7.076$, $S_{xx} = 294$. The value of the t-statistic for the hypothesis test $H_0 : b = 0$ against H_1 :

$b \neq 0$ is $t = 9.5649$ which gives a two-sided p-value of 2.386×10^{-6}. This is highly statistically significant and we have strong evidence that distance and time are highly correlated.

Also, using the table it is easy to construct confidence intervals for the slope and intercept. For example a 99% CI for the slope is $3.94728 \pm t(10, 0.005) \times 0.412684 = 3.94728 \pm 3.16927 \times 0.412684$.

The ANOVA table is

Source	DF	SS	MS	F-statistic	p-Value
Regression	1	4580.82	4580.82	91.4873	2.386×10^{-6}
Residuals	10	500.705	50.0705		
Total	11	5081.52			

We see that the p-value for the F-statistic gives the same result for the hypothesis test on the slope.

Remark 6.20   The random variable $T = \frac{\overline{X} - \mu_0}{S/\sqrt{n}}$ when we have a random sample $X_1, \ldots, X_n$ from a normal population has a Student's t-distribution with $n - 1$ degrees of freedom. Also, $\frac{n-2}{\sigma^2} S^2 \sim \chi^2(n-2)$. So now the T variable may be rewritten as

$$T = \frac{\overline{X} - \mu_0}{S/\sqrt{n}} = \frac{\frac{\overline{X} - \mu_0}{\sigma/\sqrt{n}}}{\sqrt{S^2/\sigma^2}} = \frac{Z}{\sqrt{S^2/\sigma^2}}$$

and then

$$T^2 = \frac{Z^2/1}{S^2/\sigma^2} = \frac{\chi^2(1)}{\chi^2(n-2)/(n-2)} = F(1, n-2)$$

since $Z^2 \sim \chi^2(1)$. Therefore, $T^2(n-2, \alpha/2) = F(1, n-2, \alpha)$. We have to use $\alpha/2$ for the t-distribution because of the fact that T is squared, i.e, $P(F(1, n-2, \alpha) > f) = P(|T(n-2, \alpha/2)| > t) = \alpha$. That's why the two-sided test has p-value $P(F(1, n-2) > f)$.

6.4.1   CONFIDENCE AND PREDICTION BANDS

The use of the regression line is to predict the mean response $E(Y|x)$ to a given input. However, we may want to predict the actual response (and not the mean) to the given input x. Denote the given input as x_p. We want to predict both $\mu_p = a + b\, x_p$ and we want to predict the actual response y_p to this given x_p. For instance, a pediatrician may want to know the mean height for a 7-year-old (this is a mean estimation problem) but the parent of a 5-year-old may want to predict the height of her specific child when the child becomes 7.[1]

---

[1]In general, predictions of y for a given x using a regression line are only valid in the range of the data of the x values.

We denote the predicted value of the rv Y by y_p and the estimate of $E(Y_p)$ by μ_p. In the absence of other information, the best estimates of both will be given by $y_p = \mu_p = \hat{a} + \hat{b}x_p$, the difference is that the error of these estimates are not the same. In particular, we will have a **confidence interval for μ_p, but a prediction interval (abbreviated PI) for y_p.**

Proposition 6.21   Let x_p be a particular value of the input x. The standard error for the mean value associated with x_p, namely $\mu_p = \hat{a} + \hat{b}\,x_p$ is $SD(Y_p) = \sigma\sqrt{\frac{1}{n} + \frac{(x_p - \bar{x})^2}{S_{xx}}}$. If σ is unknown and we replace σ by s, the standard error is $SD(Y_p) = s\sqrt{\frac{1}{n} + \frac{(x_p - \bar{x})^2}{S_{xx}}}$.

The $100(1 - \alpha)\%$ **Confidence Interval for the conditional mean of the response** corresponding to x_p is

$$\mu_p \pm t(n - 2, \alpha/2)s\sqrt{\frac{1}{n} + \frac{(x_p - \bar{x})^2}{S_{xx}}}.$$

A future predicted value of Y corresponding to x_p, namely $y_p = \hat{a} + \hat{b}\,x_p$ has standard error $s\sqrt{1 + \frac{1}{n} + \frac{(x_p - \bar{x})^2}{S_{xx}}}$ when σ is unknown.

The $100(1 - \alpha)\%$ **Prediction Interval for the observed value of the response** corresponding to x_p is

$$y_p \pm t(n - 2, \alpha/2)s\sqrt{1 + \frac{1}{n} + \frac{(x_p - \bar{x})^2}{S_{xx}}}.$$

We will not derive these results but simply note that the only difference between them is the additional 1 inside the square root. This makes the PI for a response wider than the CI for the mean response to reflect the additional uncertainty in predicting a single response rather than the mean response.

If we let x_p vary over the range of values for which linearity holds we obtain confidence curves and prediction curves which band the regression line. The bands are narrowest at the point on the regression line $(\bar{x}, \bar{y})$. Extrapolating the curves beyond that point results in ever widening bands and beyond the range of the observed data, linearity and the bands may not make sense. Making predictions beyond the range of the data is a bad idea in general.

Example 6.22   The following table exhibits the data for 15 students giving the time to complete a test x and the resulting score y.

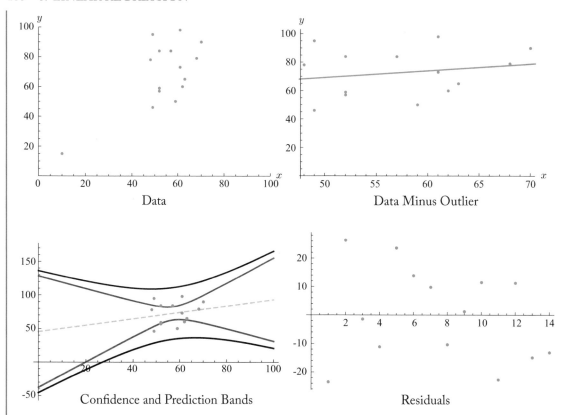

Figure 6.6: Plots for Example 6.22.

index	1	2	3	4	5	6	7	
time (x)	59	49	61	52	61	52	48	
score (y)	50	95	73	59	98	84	78	

index	8	9	10	11	12	13	14	15
time (x)	53	68	57	49	70	62	52	10
score (y)	65	79	84	46	90	60	57	15

The data has summary statistics $\overline{x} = 57.3571, \overline{y} = 72.7143, s_X = 7.17482, s_Y = 16.7398$ and $r = 0.2046$.

Figure 6.6 has a scatterplot of the data. As soon as we see the plot we see that point 15 is an outlier and is either a mistake in recording or the student gave up and quit. We need to remove this point and we will consider it dropped. Figure 6.6 shows the data points with the outlier removed and the fitted regression line. This line has equation

$$y = 45.34 + 0.4773x.$$

A plot of the data, the fitted line, and the mean confidence bands (95% confidence) and single prediction bands show that the prediction bands are much wider than the confidence bands reflecting the uncertainty in prediction. It is also clear that the use of these bands should be restricted to the range of the data.

The equations of the bands here are given by

$$y = 0.477x + 45.337 \pm 2.179\sqrt{0.435x^2 - 49.86x + 1450.66} \qquad (6.3)$$

for the 95% confidence bands, and

$$y = 0.477x + 45.337 \pm 2.179\sqrt{0.435x^2 - 49.86x + 1741.53} \qquad (6.4)$$

for the 95% prediction bands. This means that for a given input x_p =test time, the predicted test score would be $45.34 + 0.477x_p$. The 95% CI for this **mean** predicted test score is given in (6.3) and the 95% PI for this particular x_p is given in (6.4), with $x = x_p$. For example, suppose a student takes 50 minutes to complete the test. According to our linear model, we would predict a mean score of 69.2. The 95% CI for this mean score is [54.70, 83.70]. On the other hand, the 95% PI for the particular score associated with a test time=50 is [29.31, 109.1], which is a very wide interval and doesn't even make sense if the maximum score is 100.

If we test the hypothesis $H_0 : b = 0$ against the alternative $H_1 : b \neq 0$, we get the results that the p-value is $p = .4829$, which means that the null hypothesis is plausible, i.e., it may not be reasonable to predict test score from test time.

We have the ANOVA table

Source	DF	SS	MS	F-statistic	p-Value
Regression	1	152.469	152.469	0.524191	0.482936
Residuals	12	3490.39	290.866		
Total	13	3642.86			

Finally, we plot the residuals in order to determine if there is a pattern which would invalidate the model. The summary statistics for this regression tell us the correlation coefficient is $r = 0.2046$ so the coefficient of determination is $r^2 = 0.0419$ and only about 4% of the variation in the scores are explained by the linear regression line. The 95% CI for the slope and intercept are $(-.959108, 1.91375)$ and $(-37.6491, 128.322)$, respectively.

Remark 6.23   One important point of the previous example and linear regression in general is the identification and elimination of outliers. Because we are using a linear model, a single outlier can drastically affect the slope, invalidating the results. To identify possible outliers use the following method.

Using all the data points calculate the regression line and the estimate s of σ. Suppose the regression line becomes $y = a + bx$. Now consider the two lines $y = a + bx \pm 2(s)$. In other words we shift the regression line up and down by twice the SD of the residuals. Any data point lying outside this band around the regression line is considered an outlier. Remove any such points and recalculate. We use twice the SD to account for approximately 95% of reasonable data values.

6.4.2   HYPOTHESIS TEST FOR THE CORRELATION COEFFICIENT

The correlation coefficient depends on a random sample (X_i, Y_i) of size n from a joint normal distribution. The random variable giving the correlation for this random sample is $R = R(X_1, \ldots, X_n, Y_1, \ldots, Y_n)$,

$$R = \sum_{i=1}^{n} \frac{(X_i - \overline{X})}{\sqrt{\sum_{i=1}^{n}(X_i - \overline{X})^2}} \frac{(Y_i - \overline{Y})}{\sqrt{\sum_{i=1}^{n}(Y_i - \overline{Y})^2}}.$$

To analyze R we would be faced with the extremely difficult job of determining its distribution. The only case when this is not too hard is when we want to consider the hypothesis test $H_0 : \rho = 0$ because we know that this hypothesis is equivalent to testing $H_0 : b = 0$ if the slope of the regression line is zero. This is due to the formula $b = \rho \frac{\sigma_Y}{\sigma_X}$.

We will work with the sample values $(x_i, y_i), \overline{x}, \overline{y}, s_x, s_y$, and r. Now we have seen that $r^2 = \hat{b}^2 \frac{S_{xx}}{S_{yy}} = \frac{SSR}{SST} = 1 - \frac{SSE}{SST}$ so that

$$r = \hat{b}\frac{s_x}{s_y} = \hat{b}\frac{\sqrt{\frac{1}{n-1}S_{xx}}}{\sqrt{\frac{1}{n-1}S_{yy}}} = \hat{b}\sqrt{\frac{S_{xx}}{S_{yy}}} = \hat{b}\sqrt{\frac{S_{xx}}{SST}}, \text{ and } 1 - r^2 = \frac{SSE}{SST} = \frac{(n-2)s^2}{SST}.$$

We also know that the test statistic for $H_0 : b = 0$ is $t = \dfrac{\hat{b} - 0}{SE(\hat{b})}$ and is distributed as $t(n-2)$.

Since $SE(\hat{b}) = \dfrac{s}{\sqrt{S_{xx}}}$ we have

$$t = \frac{\hat{b}}{SE(\hat{b})} = \hat{b}\frac{\sqrt{S_{xx}}}{s} = \hat{b}\sqrt{\frac{S_{xx}}{SST}}\sqrt{\frac{(n-2)SST}{(n-2)s^2}}.$$

But using the formulas for r and $1 - r^2$ we have

$$\hat{b}\sqrt{\frac{S_{xx}}{SST}}\sqrt{\frac{(n-2)SST}{(n-2)s^2}} = \frac{r\sqrt{n-2}}{\sqrt{1-r^2}}.$$

We conclude that the test statistic for $H_0 : \rho = 0$ is

$$\boxed{t = \frac{r\sqrt{n-2}}{\sqrt{1-r^2}} \text{ and is distributed as } t(n-2).}$$

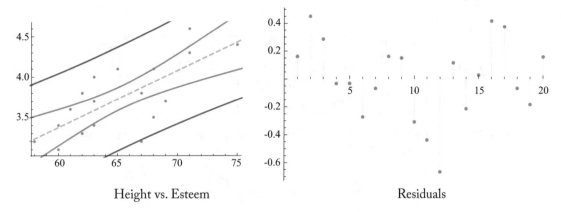

Height vs. Esteem   Residuals

Figure 6.7: Height vs. esteem.

More precisely, the rv $T = \dfrac{R\sqrt{n-2}}{\sqrt{1-R^2}} \sim t(n-2)$.

Example 6.24   Does a person's self esteem depend on their height? (See Figure 6.7.) The following table gives data from 20 people.

Person	Height	Self Esteem	Person	Height	Self Esteem
1	68	4.1	11	68	3.5
2	71	4.6	12	67	3.2
3	62	3.8	13	63	3.7
4	75	4.4	14	62	3.3
5	58	3.2	15	60	3.4
6	60	3.1	16	63	4.0
7	67	3.8	17	65	4.1
8	68	4.1	18	67	3.8
9	71	4.3	19	63	3.4
10	69	3.7	20	61	3.6

The sample correlation coefficient is $r = 0.730636$. The fitted regression equation is $y = -.866269 + 0.07066x$. The 95% CI's for the slope and intercept are $[0.0379, 0.10336]$ and $[-3.00936, 1.27682]$, respectively. The calculation for the slope and intercept are shown in the next table.

	Estimate	Standard Error	t-Statistic	p-Value
$\hat{a} = $ Intercept	−0.866269	1.02007	−0.849223	0.406911
$\hat{b} = $ Slope	0.0706616	0.0155639	4.54009	0.000253573

The ANOVA table is

Source	DF	SS	MS	F-statistic	p-Value
Regression	1	1.84144	1.84144	20.6124	0.000253573
Error	18	1.60806	0.0893366		
Total	19	3.4495			

To test the hypothesis $H_0 : \rho = 0$ against $H_1 : \rho \neq 0$, we see that the t statistic for the correlation (and also the slope) is

$$t(18) = 4.54009 = \frac{.73\sqrt{20 - 2}}{\sqrt{1 - .73^2}}$$

which results in a p-value[2] of 0.00025, i.e., $P(|t(18)| \geq 4.54009) = 0.00025$. Thus, we have high statistical significance and plenty of evidence that the correlation (and the slope) is not zero. Incidentally, the residual plot shows that there may be an outlier at person 12.

Example 6.25   Suppose we calculate $r = 0.32$ from a sample of size $n = 18$. First we perform the test $H_0 : \rho = 0, H_1 : \rho > 0$. The statistic is

$$t = \frac{r\sqrt{n - 2}}{\sqrt{1 - r^2}} = \frac{.32\sqrt{16}}{\sqrt{1 - .32^2}} = 1.35.$$

For 16 degrees of freedom we have $P(t(16) > 1.35) = 0.0979$, so we do not reject the null.

Next, we find the sample size necessary in order to conclude that $r = 0.32$ differs significantly from 0 at the level $\alpha = 0.05$ level. In order to reject $H_0 : \rho = 0$ with a two-sided alternative, we would need $|t(n - 2, 0.025)| \leq t = \frac{r\sqrt{n-2}}{\sqrt{1-r^2}}$. The sample size needed is the solution to the equation

$$t(n - 2, \ 0.05) = \frac{.32\sqrt{n - 2}}{\sqrt{1 - .32^2}}$$

for n. In general, this cannot be solved exactly. By trial and error, we have for $n = 38, t(36, 0.025) = 2.02809$ and $0.33776\sqrt{38 - 2} = 2.02656$. Also, for $n = 39, t(37, 0.025) = 2.026 < 0.33776\sqrt{37} = 2.0541$. Consequently, the first n which works is $n = 39$.

Testing $H_0 : \rho = \rho_0$ Against $H_1 : \rho \neq \rho_0$

It is possible to show that

$$Z = \frac{1}{2} \ln\left(\frac{1 + R}{1 - R}\right) \text{ with inverse } R = \frac{e^{2Z} - 1}{e^{2Z} + 1},$$

---

[2]The p-value is obtained using a TI-83 as $2\text{tcdf}(4.54009, 999, 18) = 0.0002$.

has an **approximate** normal distribution with mean $\frac{1}{2}\ln\left(\frac{1+\rho}{1-\rho}\right)$ and SD $= \frac{1}{\sqrt{n-3}}$. This means we can base a hypothesis test $H_0 : \rho = \rho_0$ on the test statistic

$$z = \frac{\frac{1}{2}\ln\frac{1+r}{1-r} - \frac{1}{2}\ln\frac{1+\rho_0}{1-\rho_0}}{\sqrt{\frac{1}{n-3}}}$$

which comes from the usual formula $z = \frac{\text{observed}-\text{expected}}{SE}$. Then we proceed to reach a conclusion based on our choice of alternative:

$$H_1 : \rho < \rho_0 \quad \text{then, reject null if} \quad z \le -z_\alpha$$
$$H_1 : \rho > \rho_0 \quad \text{then, reject null if} \quad z \ge z_\alpha$$
$$H_1 : \rho \ne \rho_0 \quad \text{then, reject null if} \quad |z| \ge z_{\alpha/2}.$$

Using the statistic z, we may also construct a $100(1-\alpha)\%$ CI for ρ by using

$$\frac{1}{2}\ln\frac{1+r}{1-r} \pm z_{\alpha/2}\frac{1}{\sqrt{n-3}}$$

and then transforming back to r. Here's an example.

Example 6.26   A sample of size $n = 20$ resulted in a sample correlation coefficient of $r = 0.626$. The 95% CI for $\frac{1}{2}\ln\frac{1+r}{1-r} = 0.7348$ is $0.7348 \pm 1.96 \times \frac{1}{\sqrt{17}} = 0.7348 \pm 0.4754$. Using the inverse transformation, we get the 95% CI for ρ given by

$$\left(\frac{e^{2\times0.2594} - 1}{e^{2\times0.2594} + 1}, \frac{e^{2\times1.2102} - 1}{e^{2\times1.2102} + 1}\right) = (0.2537, 0.8367).$$

Suppose a second random sample of size $n = 15$ results in a sample correlation coefficient of $r = 0.405$. Based on this, we would like to test $H_0 : \rho = 0.626$ against $H_1 : \rho \ne 0.626$. Since our sample correlation coefficient $0.405 \in (0.2537, 0.8367)$ we cannot reject the null and this is not sufficient evidence to conclude the correlation coefficient is not 0.626.

Suppose another random sample led to a correlation coefficient (sample size $n = 24$) of $r = 0.75$. We now want to test $H_0 : \rho = 0.626$, $H_1 : \rho > .626$. Here we calculate

$$z = \frac{\frac{1}{2}\ln\frac{1+.75}{1-.75} - \frac{1}{2}\ln\frac{1+.626}{1-.626}}{\sqrt{\frac{1}{24-3}}} = 1.0913.$$

Since $P(Z > 1.0913) = 0.1375$ we still cannot reject the null. If we had specified $\alpha = 0.05$, then $z_{0.05} = 1.645$ and since $1.645 > 1.0913$, we cannot reject the null at the 5% level of significance.

Example 6.27   This example shows that we can compare the difference of correlations for two independent random samples. All we need note is that the SE for the difference will be given by $SE = \sqrt{\frac{1}{n_1-3} + \frac{1}{n_2-3}}$.

Suppose we took two independent random samples of sizes $n_1 = 28, n_2 = 35$ and calculate the correlation coefficient of each sample to be $r_1 = .5, r_2 = .3$. We want to test $H_0 : \rho_1 = \rho_2, H_1 : \rho_1 \neq \rho_2$. The test statistic is

$$z = \frac{\frac{1}{2} \ln \frac{1+.5}{1-.5} - \frac{1}{2} \ln \frac{1+.3}{1-.3}}{\sqrt{\frac{1}{28-3} + \frac{1}{35-3}}} = 0.8985.$$

Since $P(Z > .8985) = 0.184$, the p-value is .36 and we cannot reject the null.

6.5   PROBLEMS

6.1.   Show that the equations for the minimum of $f(a, b)$ in Proposition 6.6 are

$$\frac{\partial f}{\partial a} = -2n\,(\overline{y} - a - b\overline{x}) = 0$$

$$\frac{\partial f}{\partial b} = -2 \sum_{i=1}^{n} \left(x_i y_i - a x_i - b x_i^2\right) = 0.$$

Solve the first equation for a and substitute into the second to find the formulas for $\hat{a}$ and $\hat{b}$ in Proposition 6.6.

6.2.   We have seen that if we consider x as the independent and y as the dependent variable, the regression line is $y - \overline{y} = r\frac{s_y}{s_x}(x - \overline{x})$. What is the regression line if we assume instead that y is the independent variable and x is the dependent variable? Derive the equation by minimizing $f(a, b) = \sum_{i=1}^{n}(x_i - a - b\,y_i)^2$. Find the value of f at the optimal $\hat{a}, \hat{b}$.

6.3.   If the regression line with dependent Y is $Y = a + b\,X$ and the line with dependent X is $X = c + d\,Y$, derive that $b \times d = \rho^2$. Then, given the two lines $Y = a + 0.476\,X, X = c + 1.036\,Y$, find ρ.

6.4.   If $Y_i = 1.1 + 2.5x_i + \varepsilon_i, \varepsilon_i \sim N(0, 1.7)$ is a regression model with independent errors, find

(a) The distribution of $Y_1 - Y_2$ when Y_1 corresponds to $x_1 = 3$ and Y_2 corresponds to $x_2 = 4$.

(b) $P(Y_1 > Y_2)$.

6.5. Given the data in the table, find the equation of the regression line in the form $(y - \bar{y}) = r\frac{s_y}{s_x}(x - \bar{x})$ and also with x as the dependent variable. Find the minimum value of $f(a,b) = \sum_{i=1}^{n}(x_i - a - by_i)^2$ with dependent variable x and with dependent variable y.

x	1	4	2.2	3.7	4.8	6	6.7	7.2
y	5	4.6	3.8	4.7	5.2	5.9	6	7.8

6.6. Math and verbal SAT scores at a university have the following summary statistics:

$$\overline{MSAT} = 570, \ SD_{MSAT} = 110 \text{ and } \overline{VSAT} = 610, \ SD_{VSAT} = 120,$$

Suppose the correlation coefficient is $r = 0.73$.

(a) If a student scores 690 on the MSAT, what is the prediction for the VSAT score?

(b) If a student scores 700 on the VSAT, what is the prediction for the MSAT score?

(c) If a student scores in the 77th percentile for the MSAT, what percentile will she score in the VSAT?

(d) What is the standard error of the estimate of the regression?

6.7. We have the following summary statistics for a linear regression model relating the heights of sisters and brothers:

$$\bar{B} = 68, \ SD_B = 2.4, \text{ and } \bar{S} = 62, \ SD_S = 2.2, \ n = 32.$$

The correlation coefficient is $r = 0.26$.

(a) What percentage of sisters were over 68 inches?

(b) Of the women who had brothers who were 72 inches tall, what percentage were over 68 inches tall?

6.8. Suppose the correlation between the educational levels of brothers and sisters in a city is 0.8. Both brothers and sisters averaged 12 years of school with an SD of 2 years.

(a) What is the predicted educational level of a woman whose brother has completed 18 school years?

(b) What is the predicted educational level of a brother whose sister has completed 16 years of school?

6.9. In a large biology class the correlation between midterm grades and final grades is about 0.5 for almost every semester. Suppose a student's percentile score on the midterm is

(a) 4% (b) 75% (c) 55% (d) unknown

Predict the student's final grade percentile in each case.

6.10. In many real applications it is known that the y-intercept must be zero. Derive the least squares line through the origin that minimizes $f(b) = \sum_{i=1}^{n}(y_i - bx_i)^2$ for given data points $\{(x_i, y_i)\}_{i=1}^{n}$.

6.11. Find the equations, but do not solve, for the best quadratic approximation to the data points $\{(x_i, y_i)\}_{i=1}^{n}$. That is, find the equations for a, b, c which minimize $f(a, b, c) = \sum_{i=1}^{n}(y_i - a - bx_i - cx_i^2)^2$. Now find the best quadratic approximation to the data $(-3, 7.5), (-2, 3), (-1, 0.5), (0, 1), (1, 3), (2, 6), (3, 14)$.

6.12. This problem shows how to get the estimates of a, b, and σ by maximizing a function called the likelihood function. Recall that we assume $\varepsilon = Y - a - bx \sim N(0, \sigma)$. Define, for the data points $\vec{x} = (x_1, \ldots, x_n)$ and $\vec{y} = (y_1, \ldots, y_n)$,

$$L(a, b, \sigma; \vec{x}, \vec{y}) = \prod_{i=1}^{n} \frac{1}{\sigma\sqrt{2\pi}} \exp\left(-\frac{(y_i - a - bx_i)^2}{2\sigma^2}\right).$$

This is called the **likelihood function**. It is a measure of the probability $P(Y_1 = y_1, \ldots, Y_n = y_n)$, that we obtain the sample data $(y_1, \ldots, y_n)$ from the experiment. Maximizing this function is based on the idea that we want to determine the parameters a, b, σ which make obtaining the data we did obtain as likely (or probable) as possible.

(a) Take the log of L, say $G(a, b, \sigma; \vec{x}, \vec{y}) = \ln L(a, b, \sigma; \vec{x}, \vec{y})$. Use calculus to find $\frac{\partial G}{\partial a}, \frac{\partial G}{\partial b}$, and $\frac{\partial G}{\partial \sigma}$. Set them to zero and show that the equations for a, b, and σ become

$$\frac{\partial G}{\partial a} = \frac{1}{\sigma^2}\sum_{i=1}^{n}(y_i - a - bx_i) = 0$$

$$\frac{\partial G}{\partial b} = \frac{1}{\sigma^2}\sum_{i=1}^{n}(y_i - a - bx_i)x_i = 0$$

$$\frac{\partial G}{\partial \sigma} = -\frac{1}{\sigma^3}\left(n\sigma^2 - \sum_{i=1}^{n}(y_i - a - bx_i)^2\right) = 0.$$

(b) What is the connection of the first two equations with Proposition 6.6?

(c) The third equation gives the estimator for σ^2, $s^2 = \frac{1}{n}\sum_{i}(y_i - a - bx_i)^2$. Assuming

$$\frac{1}{\sigma^2}S^2 = \frac{1}{\sigma^2}\sum_{i}(Y_i - a - bx_i)^2 \sim \chi^2(n-2),$$

find $E(S^2/\sigma^2)$ and then find an unbiased estimator of σ^2 using $s^2 = \frac{1}{n} \sum_i (y_i - a - bx_i)^2$.

6.13. In the following table we have predicted values from a regression line model and the actual data values. Calculate the residuals and the SE estimate of the regression, s.

y_i	55	64	48	49	58
$\hat{y}_i$	62	61	49	51	50

6.14. The table contains Consumer Price index data for 2008–2018 for gasoline and food

Year	08	09	10	11	12	13	14	15	16	17	18
Gas	34.5	-40.4	51.3	13.4	9.7	-1.5	0.1	-35.4	-7.3	20.3	8.5
Food	4.8	5.2	-0.2	1.8	4.2	1.6	1.1	3.1	0.9	-0.1	1.6

Find the correlation coefficient and the regression equation. Find the standard error of the estimate of σ.

6.15. In a study to determine the relationship between income and IQ, we have the following summary statistics:

mean income $= 95,000$, $SD = 38,000$, and mean IQ $= 105$, $SD = 12$, $r = 0.6$.

(a) Find the regression equation for predicting income from IQ.

(b) Find the regression equation for predicting IQ from income.

(c) If the subjects in the data are followed for a year and everyone's income goes up by 10%, find the new regression equation for predicting income from IQ.

6.16. In a sample of 10 Home Runs in Major League Baseball, the following summary statistics were computed:

	Mean	Standard Deviation
Distance	416.6	35.88
SpeedOffBat	104.26	3.26
Apex	95.3	25.74

We are also given the correlation data that the correlation between SpeedOffBat and Distance is 0.098, the correlation between the Apex and Distance is -0.058, and the correlation between SpeedOffBat and Apex is 0.3977.

(a) Find the regression lines for predicting distance from SpeedOffBat and from Apex.

(b) Find the ANOVA tables for each case.

(c) Test the hypotheses that the correlation coefficients in each case are zero.

6.17. The duration an ulcer lasts for the grade of ulcer is given in the table.

Stage(x)	4	3	5	4	4	3	3	4	6	3
Days(y)	18	6	20	15	16	15	10	18	26	15
Stage(x)	3	4	3	2	3	2	2	3	5	6
Days(y)	8	16	17	6	7	7	8	11	21	24

Find the ANOVA table and test the hypothesis that the slope of the regression line is zero.

6.18. Consider the data:

x	-1	0	2	-2	5	6	8	11	12	-3
y	-5	-4	2	-7	6	9	13	21	20	-9

Find the regression line and test the hypothesis that the slope of the line is zero.

6.19. The following table gives the speed of a car, x, and the stopping distance, y, when the brakes are applied with full force.

x km/h	10	30	50	70	90	110	120
y meters	0.7	6.2	17.2	33.8	55.8	83.4	99.2

Fit a regression line to this data and plot the line and the data points on the same plot. Also plot the residuals. Test the hypothesis that the data is uncorrelated.

6.20. In the hypothesis testing chapter we considered testing $H_0 : \mu_1 = \mu_2$ from two independent random samples. We can set this up as a linear regression problem using the following steps.

- The sample sizes are n_1 and n_2. The y-data values are the observations from the two samples, labeled $y_1, \ldots, y_{n_1}, y_{n_1+1}, \ldots, y_{n_1+n_2}$. The x-data values are defined by

$$x_i = \begin{cases} 1 & \text{if } y_i \text{ comes from sample 1} \\ 0 & \text{if } y_i \text{ comes from sample 2} \end{cases}, \quad i = 1, 2, \ldots, n_1 + n_2.$$

- Calculate the regression line for the data values $\{(x_i, y_i)\}_{i=1}^{n_1+n_2}$.

(a) Use the formulas for $\hat{a}, \hat{b}$ and show that $\hat{a} = \bar{y}_2$ and $\hat{b} = \bar{y}_1 - \bar{y}_2$. Here, $\bar{y}_1$ is the sample mean for sample 1 and $\bar{y}_2$ is the sample mean for sample 2.

(b) Show that MSE for regression is the same as the pooled estimate s^2 of σ^2 with $n_1 + n_2 - 2$ degrees of freedom.

(c) Show that the regression t-test of $H_0 : b = 0$ is the same as the pooled variances t-test of $H_0 : \mu_1 = \mu_2$.

6.21. Given the following data set

x	-6	-2	2	2	4
y	12	8	6	2	2

solve the following without the use of a calculator or computer.

(a) Plot the scatter diagram, and indicate whether x, y appear linearly related.

(b) Find the regression equation for the data.

(c) Plot the regression equation and the data on the same graph. Does the line appear to provide a good fit for the data points?

(d) Compute SSE and s^2.

(e) Estimate the expected value of Y when $x = -1$.

(f) Find the correlation coefficient r and the coefficient of determination.

6.22. A study of middle- to upper-level managers is undertaken to investigate the relationship between salary level, Y, and years of work experience, X. A random sample sample of 20 managers is chosen with the following results (in thousands of dollars):

$$\sum x_i = 235, \sum y_i = 763, S_{xx} = 485.75, S_{yy} = 2236.1, S_{xy} = 886.85.$$

It is further assumed that the relationship is linear.

(a) Find $\hat{a}, \hat{b}$ and the estimated regression equation.

(b) Find the correlation coefficient, r.

(c) Find r^2 and interpret it value.

(d) Suppose the errors have distribution $\varepsilon \sim N(0, 50)$. Find $P(Y > 100)$ when $x = 12$.

6.23. An analysis of family income and energy use produced the following table from a random sample of 25 families:

	Value	SE	t-statistic	p-Value
Intercept	82.036	2.054	39.94	0
Slope	0.93051	0.05727	16.25	0
	$r^2 = .92$			

The ANOVA table is

Source	DF	SS	MS	F-statistic	p-Value
Regression			7,626.6	264.02	0
Error	23				
Total		8,291			

(a) Fill in all the missing entries.

(b) Find the regression equation.

(c) Is there sufficient evidence to conclude that X and Y are linearly related? Do the hypothesis test.

(d) Find a confidence interval and a prediction interval for families with an annual income of $40,000. Assume that $\bar{x} = 35,000$ and $s_X^2 = 3,000$.

6.24. You want to predict the cardiac output for a level of exercise of 750 kg-m per minute. The regression line is $y = 4.97 + 0.0133x$ with data statistics $s = 0.68$, $\sum x_i = 9000$, $\sum y_i = 219.5$, $\sum x_i^2 = 6,300,000$, $\sum y_i^2 = 2812.05$, and $\sum x_i y_i = 128,790$. There are 20 data points. Find the predicted cardiac output and find a 95% prediction interval. The value $x = 750$ is within the range of the data points.

6.25. Given the following summary statistics of data

$$n = 17, \sum x = 660, \sum x^2 = 35990,$$
$$\sum y = 5712, \sum y^2 = 2243266, \sum xy = 188429,$$

calculate $S_{xx}, S_{xy}, S_{yy}, \hat{b}, \hat{a}$, and s^2.

APPENDIX A

Answers to Problems

A.1   ANSWERS TO CHAPTER 1 PROBLEMS

1.1 $p = 0.3$ and $p = 3/7$.

1.2   (a) $P(A \cap B) = 1/12$.   (b) $P(A^c \cup B) = 9/12$.

1.4   (a) Let C = cigarettes, C_g = cigars. Then $P(C^c \cap C_g^c) = 0.64$.
(b) $P(C_g \cap C^c) = .04$.

1.5   (a) $A \cap B^c \cap C^c$.

       (b) $A \cap C \cap B^c$

       (c) $A \cup B \cup C$

       (d) $(A \cap B) \cup (A \cap C) \cup (B \cap C)$

       (e) $A \cap B \cap C$

       (f) $A^c \cap B^c \cap C^c = (A \cup B \cup C)^c$

       (g) $(A \cup B \cup C) \cap (A \cap B \cap C)^c$

       (h) $(A \cap B^c \cap C^c) \cup (A^c \cap B \cap C^c) \cup (A^c \cap B^c \cap C) \cup (A^c \cap B^c \cap C^c)$ or $((A \cap B) \cup (B \cap C) \cup (A \cap C))^c$

       (i) $((A \cap B) \cap C^c) \cup ((A \cap C) \cap B^c) \cup ((B \cap C) \cap A^c)$

       (j) $A \cup B \cup C \cup (A^c B^c C^c) = S$.

1.7 $P(A) = 2/3$.

1.9   Solve $\binom{x-2}{2} = 2\binom{x-2}{4}$ to get $x = 7$.

1.10 $P(C^c \cap D) = P(D) - P(C \cap D) = 0.2$.

1.11   Call the two outcomes C_1, C_2. Then $p = (\sqrt{5} - 1)/2 = 0.618$.

1.12 $P(\text{1st H on toss 5}) = (1 - p)^4 p$. $P(\text{5 tosses to 2 H's}) = 4(1 - p)^3 p^2$.

1.14 $P(A) = \dfrac{3\pi}{9\pi}$.

1.16 $P(A \cap B) \geq 0.9 + 0.9 - 1 = 0.8$.

1.17 $P(\text{other toss is H} \mid \text{one is H}) = 1/3$.

1.18   This hand has the pattern AABBC where A, B, and C are from distinct kinds. The probability is 0.047539.

1.19   One way to do this is with truth tables. For example:

A	B	$A \cap B$	$(A \cap B)^c$	A^c	B^c	$A^c \cup B^c$
0	0	0	1	1	1	1
0	1	0	1	1	0	1
1	0	0	1	0	1	1
1	1	1	0	0	0	0

In the table 0 indicates an outcome is not in the event and a 1 indicates it is. Notice that all possibilities are covered in this table and that the column $(A \cap B)^c$ and $A^c \cup B^c$ are identical. Therefore, the two events must be the same.

$$P(A^c \cap B^c) = 1 - P(A \cup B) = 1 - P(A) - P(B) + P(A \cap B),$$
$$P((A^c \cap B) \cup (A \cap B^c)) = P(A^c \cap B) + P(A \cap B^c) = P(B) + P(A) - 2P(A \cap B).$$

1.21 The answer is no.

1.22 $P(H) = 1/2$.

1.23 (a) $P(C_2 > C_1) = .47058$.

(b) $P(C_1 = C_2) = 3/51$.

1.24 Let $W = \{$team wins the game$\}$. $P(W) = 0.25$.

1.25 $P($coin is HH$|5$ H's$) = 0.8$.

1.26 (a) $P(C) = 1/2$. The events are pairwise independent.

(b) A, B, C are not mutually independent.

1.27 (a) $P(A \cup B) = 1/2$.

(b) $P(B) = 1/3$.

1.28 $P(A^c \cap B \cap C) = 1/8$.

1.29 $P(B) = 1/2$.

1.30 $P($Win$) = 2/7$.

1.31 (a) $P(D|T) = 0.2$.

(b) $P(D|TTT) = \dfrac{0.729(.04)}{0.729(.04) + .003375(.96)} = 0.9$.

1.32 $P($at least 1 6$) = 11/36$, $P($at least 1 6$|$faces different$) = 10/36$.

1.33 $P(NS|D) = \dfrac{1}{4}$.

1.34 (i) $P(A \cap B \cap C) = .5(.8).9$, $P(A \cap B \cap C^c \cup A \cap B^c \cap C \cup A^c \cap B \cap C) = .04 + .09 + .36$, $P(A^c B^c C^c) = .5(.2).1$; (ii) $P(A \cap B \cap C^2) + P(a \cap C \cap B^c) + P(B \cap C \cap A^c)$; (iii) $P((A \cup B \cup C)^c) = P(A^c)P(B^c)P(C^c)$.

1.35 Let x be the number of red balls in the second box, $x = 11$.

1.37 (a) $P(7H) = \binom{10}{7}0.4^70.6^3\frac{1}{2} + \binom{10}{7}0.7^70.3^3$.

(b) Let A =first toss is H, then $P(7H|A) = \binom{9}{6}.4^6.6^3\frac{4}{11} + \binom{9}{6}.7^6.3^3\frac{7}{11}$.

1.38 $P(A|B) = \sum_i P(A \cap E_i|B) = \sum_i \dfrac{P(A \cap B \cap E_i)}{P(B \cap E_i)}\dfrac{P(B \cap E_i)}{P(B)}$.

1.39 If we calculate the proportion of male math majors at each university, we get 0.2 at university 1 and 0.3 at university 2. For females it is $0.15 < 0.2$ at university 1 and $0.25 < 0.3$ at university 2. These inequalities are reversed in the amalgamated table.

Amalgamated	Math Major	Other	Total	Proportion
Males	230	370	1100	0.209
Females	1150	3850	5000	0.23

1.40 Let R = recover, D = die. We have $P(R|M \cap D) = .27 < P(R|M \cap D^c) = .33$ and $P(R|F \cap D) = .642 < P(R|F \cap D^c) = .66$ so the recovery probability is lower for both males and females on the drug. However, $P(R|) \cap D) = 0.538?P(R|O \cap D^c) = .44$ so that amalgamated, the drug is better,

A.2   ANSWERS TO CHAPTER 2 PROBLEMS

2.1 $P(X = 0) = \frac{6}{36}$, $P(X = 1) = \frac{10}{36}$, $P(X = 2) = \frac{8}{36}$, $P(X = 3) = \frac{6}{36}$, $P(X = 4) = \frac{4}{36}$, $P(X = 5) = \frac{2}{36}$. Then $P(0 < X \le 3) = \frac{24}{36}$. Also, $P(1 \le X < 3) = \frac{18}{36}$.

2.2 (a) $P(X = 1) = 1/4$, $P(X = 2) = 1/6$, $P(X = 3) = 1/12$.

(b) $P(1/2 < X < 3/2) = F(3/2) - F(1/2) = 1/2$.

2.3 (a) $P(e^X \le x) = P(X \le \ln x) = F_X(\ln x)$ if $x > 0$ and 0 otherwise.

(b) $P(aX + b \le x) = P(X \le \frac{x-b}{a}) = F_X((x - b)/a)$ if $a > 0$, $P(aX + b \le x) = P(X \ge (x - b)/a) = 1 - F_X((x - b)/a) + P(X = \frac{x-b}{a})$ if $a < 0$.

2.4 (a) $c = \frac{n(n+1)}{2}$.

(b) $c = 2$.

2.5   (a) $F(x) = \begin{cases} 0, & x \leq 0 \\ 3/4x, & 0 \leq x \leq 1, \\ 3/4, & 1 \leq x \leq 2, \\ 3/4 + 1/4(x-2), & 2 \leq x \leq 3 \\ 1, & x > 3. \end{cases}$

2.7   (a) $c = 1.$

(b) $F_X(x) = \begin{cases} 0, & x \leq -3 \\ \dfrac{1}{2}(x+3)^2, & -3 < x \leq -2 \\ \dfrac{1}{2}, & -2 \leq x \leq 2 \\ 1 - \dfrac{1}{2}(3-x)^2, & 2 < x \leq 3 \\ 1, & x > 3. \end{cases}$

2.8   (a) $P(X \leq 0.55) = 0.595.$

(b) $\int_0^c f(x)\,dx = 0.5 \implies c = 0.5.$

(c) $\int_0^c f(x)\,dx = 0.75 \implies c = 0.646447.$

2.9   Using a calculator we have $P(X \leq c) = 0.9$ and $c = \text{invNorm}(0.9, 12, 4) = 17.1262.$

2.10   0.364577.

2.11 $P(X \leq 70) = \text{normalcdf}(0, 70.5, 75, \sqrt{100(.75)(.25)}) = 0.149348$ using the normal approximation. The exact value using the binomial is $P(X \leq 70) = \text{binomcdf}(100, 0.75, 70) = 0.149541.$

2.12   Using the binomial distribution we have (a) $P(X = 3) = \text{binompdf}(2000, 0.001, 3) = 0.180537.$ (b) $P(X > 2) = 1 - P(X \leq 1) = 1 - \text{binomcdf}(2000, 0.001, 1) = 0.59412.$ Using the normal distribution, (a) $P(X = 3) = \text{normalcdf}(2.5, 3.5, 2, 1.4135) = 0.217469$ (b) $P(X > 2) = \text{normalcdf}(2.5, \infty, 2, 1.4135) = 0.36177.$ Using the Poisson approximation (a) $P(X = 3) = \text{poissonpdf}(2, 3) = 0.180447.$ (b) $P(X > 2) = 1 - P(X \leq 1) = 1 - \text{poissoncdf}(2, 1) = 0.59399.$ Poisson is a much better approximation but note that $n \cdot p = 2000(0.001) = 2 < 5$, so there is no real justification for using the normal approximation anyway.

2.13   Let X_i be the number of occurrences of each outcome in a single game, $i = 1, 2, \ldots, 6$.

$$P\,(X_1 = 1, X_2 = 2, X_3 = 0, X_4 = 2, X_5 = 0, X_6 = 0)$$

$$= \binom{5}{1, 2, 0, 2, 0, 0}(0.662)^1\,(0.052)^2(0.213)^0(0.018\,)^2(0.009\,)^0(0.046\,)^0$$

$$= 0.0000174.$$

2.14 $\dfrac{(r-k)(n-k)}{(k+1)(N-r-n+k+1)}.$

2.15   (a) $X \sim \text{Binom}(8, 0.1) \implies P(X = 2) = 0.1488$, $Y \sim \text{Poisson}(0.8) \implies P(Y = 2) = 0.143785$.

(b) $X \sim \text{Binom}(10, 0.95) \implies P(X = 9) = 0.315125$, $Y \sim \text{Poisson}(9.5) \implies P(Y = 9) = 0.130003$.

2.16   (a) $P(X = 1) = \text{binompdf}(50, 0.01, 1) = 0.305559$.

(b) $P(X \geq 1) = 0.394994$.

(c) $P(X \geq 2) = 0.0894353$.

2.19 $P(1/2 < X \leq 3/4) = 5/16$. The pdf of X is $f(x) = 2x, 0 < x < 1$, and 0 otherwise.

2.20   (a)   The largest area possible is $\frac{1}{2}$. ; $\{(x, y) \mid x \in [2, 3], y \in [1, 3/2]\}$.

(b) $F(a) = P(A \leq a) = P(h \leq 2a) = 2a - 1, \frac{1}{2} \leq a \leq 1$.

(c) $f(a) = \begin{cases} 2, & \dfrac{1}{2} \leq a \leq 1 \\ 0 & \text{otherwise.} \end{cases}$

2.22   (a) $P(X < 4.5) = 0.25$.

(b)   0.

(c)   continuous.

2.23 $P(X > 5) = 0.3678$.

2.24 $m = -1/0.2 \ln(0.5) = 3.46574$.

2.25 $P(Z \leq 1.28155) = 0.9$, $P(-1.64485 < Z < 1.64485) = 0.9$.

2.26   (a) $P(X > 2) = 0.090718$.

(b) $P(X > 10 | X > 9) = P(X > 1) = 0.301194$.

2.27 $P(X > 8) = e^{-4/9}$.

2.28 3333.33.

2.30 (a) Denote them by SA and SB. Then $E(SA) = 200$, $SD(SA) = 300$ and $E(SB) = 225$, $SD(SB) = 736.122$. The coefficient of variations are $C_A = 3/2 = 1.5$, $C_B = 736.122/225 = 3.27$. Stock A is a better deal.

(b) The coefficient of variation measure how much risk (measured by SD) there is relative to the amount of expected return measured by μ.

(c) A is still better because $Eg(A) = 18.228$, $SD(g(A)) = 17.852$, $C(g(A)) = 8.2304/18.228 = 0.4515$ and $Eg(B) = 10.31$, $SD(g(B)) = 17.852$, $C(g(B)) = 17.852/10.31 = 1.7315$. A still has less risk relative to the return expected.

2.31 (a) $P(X < 1/2) = \frac{1}{4}$. $P(1/4 < X \leq 1/2) = 3/16$. $P(X < 3/4 | X > 1/2) = 5/12$.

(b) $EX = \frac{2}{3}$, $SD(X) = 0.236$, $E[e^{tX}] = \dfrac{2 + 2e^t(t-1)}{t^2}$, $t \neq 0$.

2.32 (a) $EX = \frac{1}{5}$, $Var[X] = 25$, $med[X] = \frac{1}{5}\ln 2$.

2.33 $M(t) = \dfrac{p}{1-p}\dfrac{e^t(1-p)}{1-e^t(1-p)}$, $e^t(1-p) < 1$, $t < -\ln(1-p)$

$EX = \dfrac{1}{p}$ $EX^2 = \dfrac{2-p}{p^2}$, $Var(X) = \dfrac{1-p}{p^2}$.

2.34 $M(t) = e^{\lambda(e^t - 1)}$, $EX = \lambda$, $Var(X) = \lambda$.

2.35 (a) $P(X = -2) = .09$, $P(X = -1) = .24$, $P(X = 0) = .34$, $P(X = 1) = .24$, $P(X = 2) = .09$., $P(X \leq 0) = 0.67$.

(b) $EX = 0$.

2.36 $9/2$.

2.37 $MD(X) = \dfrac{35}{18}$.

2.38 When $f_X(x) = \lambda e^{-\lambda x}$, $x > 0$, we have $MD(X) = \frac{2}{\lambda e}$. When $f_X(x) = \frac{1}{b-a}$, $a \leq x \leq b$, we have $MD(X) = \frac{b-a}{4}$.

2.39 (a) 0.32.

(b) $x_{.85} = 72.614$.

2.40 (a) $EX = \dfrac{n+1}{2}$. $Var(X) = \dfrac{n^2 - 1}{12}$.

(b) $EX = \dfrac{r}{r+1}$. $Var(X) = \dfrac{r}{(r+1)^2(r+2)}$.

2.41 $M_N(t) = \left(\dfrac{pe^t}{1 - e^t(1-p)}\right)^r$. $EN = \dfrac{r}{p}$, and $Var(N) = \dfrac{r(1-p)}{p^2}$.

2.42 Normal with mean $np = n\dfrac{k}{N}$, and SD $= \sqrt{np(1-p)}\dfrac{N-n}{N-1}$.

2.43 (a) $P(X=2) = p^2$. $P(X=3) = 2p^2(1-p)$. $P(X=4) = 3p^2(1-p)^2$.

(b) There are 2 H's with the last toss resulting in a H.

(c) $EX = \dfrac{2}{p}$.

2.44 $2/3$.

2.45 $E(2+X)^2 = 14$. and $Var(4+3X) = 45$.

2.46 (a) $P(X \ge 2) = 0.0351$.

(b) $P(X \le 1) = 0.9649$.

(c) 0.29166.

2.47 $P(X=0) = 6\binom{6}{0}0\ 94\binom{94}{10}10/100\binom{100}{10}10$. $EX = 10(.06)$, $Var(X) = .06(.94)10\,\dfrac{100-10}{999}$.

2.48 $c = \frac{2}{49}$. $Y = $ total amount the insurance company has to pay out, $EY = \dfrac{71}{490}$.

2.49 $EX = 2$. $E[X(X-1)] = 3$.

2.50 (a) $a = A/2$.

(b) $\lambda = \frac{1}{a}\ln 2$.

2.51 $\sigma^2 = 22.58845$.

2.52

$$f_X(x) = \begin{cases} x + \dfrac{1}{2} & 0 \le x \le 1, \\ 0 & \text{otherwise} \end{cases} \qquad f_Y(y) = \begin{cases} y + \dfrac{1}{2} & 0 \le y \le 1, \\ 0 & \text{otherwise.} \end{cases}$$

The rvs are not independent. $E(X+Y) = \frac{7}{6}$.

2.53 $P(X < 5) \approx $ normalcdf$(0.4.5, 8, 2.5922) = 0.08746$. The exact value is $P(X < 5) = $ binomcdf$(50, 0.16, 4) = 0.08078$.

2.54 Take $k = 2$ in $P(|X - \mu| < 2\sigma) \ge 1 - \frac{1}{4} = \frac{3}{4}$.

2.55 Chebychev gives $P(|X - 0| \ge 2) \le \frac{1}{4}$. The exact probability is $P(|X| \ge 2) = P(X = 2) + P(X = -2) = \frac{1}{4}$.

2.56 $P(X = 1) = 0.26695$.

A.3   ANSWERS TO CHAPTER 3 PROBLEMS

3.1   (a) 25 samples of size 2.

Sample mean	0	1/2	1	3/2	2	5/2	3	7/2	4
Probability	0.04	0.08	0.12	0.16	0.20	0.16	0.12	0.08	0.04

(b) $E(\overline{X}) = 2.$ $Var(\overline{X}) = 1.$

3.2 $E(\overline{X}) = 15/5$ and $SD(\overline{X}) = 2.059/\sqrt{2}.$ Without replacement, $SD(\overline{X}) = 2.059/\sqrt{2}\sqrt{3/4}.$

3.3   We have $\mu = 70, \sigma^2 = 540.$ Also, $P(\overline{X} = x) = 0.04, 0.14, 0.1225, 0.18, 0.315, 0.2025$ for $x = 40, 47.5, 55, 67.5, 75, 95,$ respectively. Using this distribution calculate $E(\overline{X}) = 70 = \mu$ and $Var(\overline{X}) = 270 = \sigma^2/2.$

3.4   (a) $E(\overline{X}) = 6$ and $SD(\overline{X}) = 1.15.$ (b) $\mu = 6, \sigma = 1.15\sqrt{9} = 3.45.$

3.5   (a) The mean number of defects in a sample of 2 is $2\frac{5}{3}.$ SE for the total number of defects in a sample of size 2 is $\sqrt{2} \times 1.491 \times .8944 = 1.886.$

(b) $\dfrac{1}{15}.$

3.6   (a) $E(S_{25}) = 2637.5, Var(Sum) = 3025, SD(Sum) = 55.$ We ignore the correction factor since $\sqrt{(1000 - 25)/999} = 0.9879.$

(b) $E(\overline{X}) = 105.5, SD(\overline{X}) = \sigma/\sqrt{25} = 11/5 = 2.2.$

(c) $P(98 \leq \overline{X} \leq 104) = 0.24735.$ Therefore, if $N =$ Number of sample means in this range, $N \sim$ Binomial$(150, 0.24735)$ and $E(N) = 37.102.$

(d) $P(\overline{X} < 97.5) = 0.0000138$ so the expected number of sample means less than 97.5 will be approximately 0.

3.7 $P(X_1 + \cdots + X_{50} < 10) \approx 0.1444.$

3.8   (a) $P(4.4 < \overline{X} < 5.2) = 0.6898.$

(b) 85th percentile $= 5.345.$

(c) $P(X > 7) = 0.022.$

3.9 $P(X \leq 8) \approx$ normcdf$(0, 8.5, 16, 3.66) = 0.0202.$

3.10   (a) $A \approx N(25, 10/\sqrt{50}), B \approx N(25, 10/\sqrt{100}).$

(b) $P(19 \leq A \leq 26) = 0.7602$ and $P(19 \leq B \leq 26) = 0.8413.$

3.11 $n = 25.$

3.12 $P(|\overline{X} - 5| \leq 0.5) \approx \text{normalcdf}(4.5, 5.5, 5, 1) = 0.3829$.

3.13   (a) $P(X \leq n/2) \approx \text{normalcdf}(0, np + 0.5, np, \sqrt{np(1 - p)})$.
$P(X = n/2) \approx \text{normalcdf}(np - 0.5, np + 0.5, np, \sqrt{np(1 - p)})$.
Here's the table.

Tosses n	$P(X \leq n/2)$ Exact	$P(X \leq n/2)$ Approx.	$P(X = n/2)$ Exact	$P(X = n/2)$ Approx.
10	0.623	0.6233	0.2461	0.2482
20	0.5881	0.5885	0.1762	0.1769
40	0.5627	0.5628	0.1254	0.1256
60	0.5513	0.5514	0.1026	0.1027

(b) When $n = 4$, $p = 0.5$ we have $P(X = 2) = 0.375$ exactly and $P(X = 2) \approx$ normalcdf$(1.5, 2.5, 2, 1) = 0.3829$.

3.14   The exact answer is $P(X \geq 10) = 0.0713$. The approximate answer is $P(X \geq 10) \approx 0.0667$.

3.15   (a) Approximately   normalcdf$(47.5, 72.5, 60, 5.477) = 0.9775$ or   another   way normalcdf$(0.4, 0.6, 0.5, 5.477/120) = 0.9715$ using the proportions. The exact number is binomcdf$(120, 0.5, 72) - \text{binomcdf}(120, 0.5, 47) = 0.9779$. Therefore, if 500 people do this, we expect about 489 people to get between 40 and 60% heads.

(b) The   chance   we   would   get   453   people   (or   less)   is $P(X \leq 453) \approx$ normalcdf$(0, 453, 488.75, 3.316) \approx 0$.

3.16   0.0241.

3.17   (a) $\boxed{\boxed{-1} \cdots \boxed{-1}, \boxed{+8} \cdots \boxed{+8}}$, with $4 + 8$s and $34 - 1$s.

(b) $\mu = -0.0526$, $\sigma = 2.76203$.

(c) $P(N \geq 4) = 0.266$.

(d) $P(W \geq 0) \approx 0.46207$.

3.19   (a) 0.9744.   (b) 0.09857.   (c) 0.88958.   (d) -1.782.

3.20   (a) $EY = 8$ and $Var(Y) = 16$.

(b) $P(Y > 15.507) = 0.05$, $P(Y < 3.489) = 0.1$, $P(Y < 13.361) = 0.9$, $P(Y > 2.733) = 0.95$.

(c) $P(3.489 < Y < 13.361) = 0.8$.

(d) $P(\chi^2(1) < 0.0855) = 0.23$,   and $P(Z^2 < b) = P(-\sqrt{b} < Z < \sqrt{b}) = 0.23 \implies \sqrt{b} = 0.29237$. Then $b = 0.0855$.

3.21 $P(\hat{P} \geq 0.5325) \approx \text{normalcdf}(0.5325, \infty, 0.51, 0.02499) = 0.1839.$

3.22 $P(|t_9| > 3.1622) = 0.00575.$

3.23 With replacement: $P(t_{79} \leq -5.96) = 3.34 \times 10^{-8}$, i.e., virtually no chance. Without replacement: $P(t_{79} \leq -6.95) \approx 0.$

3.24 $P(\overline{X} \geq 1200) = 0.315427.$

3.25 $P(.45 \leq \hat{P} \leq .48) = 0.654612.$

3.26 Sample 1: $\hat{p} = 0.2, SE = 0.0632.$ Sample 2: $\hat{p} = 0.25, SE = 0.0684.$ Sample 3: $\hat{p} = 0.325, SE = 0.0740.$

3.27 $E\hat{\mu} = \mu, Var(\hat{\mu}) = \frac{1}{4}2\sigma^2 = \frac{\sigma^2}{2}.$

3.28 $P(|\overline{X}| \leq 0.01) = 0.52050.$

3.29 $n \geq 2075.$

3.30 (a) $tcdf(-1.923, 1.923, 99) = 0.94265.$ (b) $n \geq 152.$

3.31 $P(t_{48} \geq 2.25/(6/7)) = 0.00579.$

3.32 (a) $P(\overline{X}_{125} - \overline{Y}_{125} \geq 160) = 0.9772.$ (b) $P(\overline{X}_{125} - \overline{Y}_{125} \geq 250) = 0.0062.$

3.33 $P(P_A - P_B \geq 0.1) = 0.15865.$

3.34 (a) 0.05826.   (b) 0.1952033.

3.35 0.071349.

3.36 (a) 0.02275.   (c) 0.453.   (e) $Y \sim \chi^2(80).$
(b) 0.3085.   (d) $c = 1.667.$

A.4   ANSWERS TO CHAPTER 4 PROBLEMS

4.1 (a) 0.47, 0.0499.   (c) T.   (e) Doesn't make sense.
(b) T.   (d) (0.37218, 0.56782).   (f) 0.47866.

4.2 (a) The 90% CI is (66.964, 67.936).   (b) (67.01, 67.89).

4.3 (a) 47.5.

(b) $X \sim \text{Binom}(50, 0.95)$. $P(X = 40) \approx 0$, $P(X \leq 45) = 0.1036$, $P(X > 40) \approx 1$.

4.4 $n \geq (1.96/0.03)^2/4 = 1067.11$.

4.5 $(0.42185, 0.67815)$.

4.6 (a) $(14.041, 16.359)$, $(13.767, 16.633)$, $(13.361, 17.039)$.

   (b) $(14.682, 15.718)$, $(14.559, 15.841)$, $(14.378, 16.022)$.

4.7 The pivot is $T = \dfrac{\overline{X}-\mu}{s_X/\sqrt{n}}$ and we start with $P(T \leq t(n-1, \alpha/2)) = 1 - \alpha$. Solving the inequality for μ to get the result.

4.8 (a) $P(X_{\min} \leq m \leq X_{\max}) = 1 - P(X_{\min} > m) - P(X_{\max} < m)$. Now $P(X_{\min} > m) = P(X_1 > m, \ldots, X_n > m) = P(X > m)^n = (1/2)^n$, and $P(X_{\max} < m) = P(X_1 < m, \ldots, X_n < m) = P(X < m)^n = (1/2)^n$, $P(X_{\min} \leq m \leq X_{\max}) = 1 - 2(1/2)^n = 1 - (1/2)^{n-1}$.

   (b) By the first part this is $1 - (1/2)^7 = 0.9921875$.

4.9 (a) $(11.15, 12.05)$. (b) (b). (c) 1138.

4.10 $n \geq 884$.

4.11 95% CI is $(37.874, 41.826)$. The histogram of the data is skewed right because there are some cars with high mpg. A lower 95% CI is $(39.85 - 1.729\frac{4.221187}{\sqrt{20}}, \infty) = (38.218, \infty)$. We are 95% confident that the mean mpg is at least 38.218 mpg.

4.12 $\left(\dfrac{(n-1)s_X^2}{\chi^2(n-1, \alpha/2)}, \dfrac{(n-1)s_X^2}{\chi^2(n-1, 1-\alpha/2)}\right) = (34 \times 4/51.9659, 34 \times 4/19.80625) = (2.6171, 6.8665)$.

4.13 (a) $(23.722, 24.278)$. (b) $(17.568, 30.432)$.

4.14 The mean difference is 4.2 pounds; the CI is $(-1.721, 10.121)$ with $df = 60.292$. There is not enough evidence to conclude the difference is real.

4.15 95% CI is $(1.7134, 2.0866)$ with $df = 284.865$. 99% CI is $(1.654, 2.146)$. Since 0 is not in either CI, there is evidence that the difference is real.

4.16 The 99% CI for σ is $s \pm 3\frac{s}{\sqrt{2n}}$. The percentage error in the SD is $\frac{3s/\sqrt{2n}}{s} = 300\frac{1}{\sqrt{2n}}\%$. If we want this no more than 5% we need $300\frac{1}{\sqrt{2n}} \leq 5 \implies n \geq 1800$.

4.17   (a) $(4.057, 8.543)$.   (b) $n \geq 346$.

4.18   (a) $(20.122, 21.478)$.   (b) $(12.473, 29.126)$.

4.19 $\overline{d} = 3.7, s = 4.945$, CI is $(0.16238, 7.238)$.

4.20   The CI is $(-0.0263, 0.22626)$.

A.5   ANSWERS TO CHAPTER 5 PROBLEMS

5.1   (a) $z = -1.0$. Retain H_0.

   (b) $\beta(2) = 0.2946$.

   (c) $\beta(\mu) = P(5.54 - 5\mu < Z < 9.46 - 5\mu)$. The power function is then $\pi(\mu) = 1 - P(5.54 - 5\mu < Z < 9.46 - 5\mu)$, and $\pi(1.5) = 0.05$.

5.2 $\alpha = 0.03$.

5.3   (a) The p-value is $= 0.054$. Retain H_0.   (c) The p-value is $= 0.032$. Reject H_0.

   (b) The p-value is $= 0.121$. Retain H_0.

5.4   (a) Reject H_0 if $t = \dfrac{105.6 - 100.3}{6.25/\sqrt{15}} = 3.2843 \geq t(14, 0.01) = 2.624$. Reject H_0.

   (b) $\beta(103) = 0.8211 \Rightarrow \pi(103) = 1 - \beta(103) = 1 - 0.8211 = 0.1789$.

   (c) The general form of $\pi(\mu)$ is $\pi(\mu) = 1 - P(t(14) < 64.778 - 0.61968\mu)$.

5.5   (a) Reject H_0 if $t \geq t(24, 0.01) = 2.492$.   (c) Reject H_0 if $t \leq -t(24, 0.025) = -2.064$ or $t \geq t(24, 0.025) = 2.064$.

   (b) Reject H_0 if $t \leq -t(24, 0.02) = -2.172$.

5.6   (a) p-value $= 0.0176$. Reject H_0.   (c) p-value $= 2P(t(19) \leq -1.1849) = 2(0.1253) = 0.2506$. Retain H_0.

   (b) p-value $= 0.234$. Retain H_0.

5.7   (a) We have $\alpha = P\left((n-1)S^2/\sigma_0^2 \leq \chi^2(n-1, 1-\alpha)\right)$. Therefore by the definition of Type II error,

$$\beta(\sigma_1^2) = P\left((n-1)S^2/\sigma_1^2 > \sigma_0^2/\sigma_1^2 \chi^2(n-1, 1-\alpha)\right).$$

The remaining parts are similar.

5.8　(a) $\chi^2 = \frac{(n-1)s^2}{\sigma_0^2} = \frac{19(0.33167)}{1^2} =$
6.3017. Reject H_0 if $\chi^2 = 6.3017 \geq$
$\chi^2(19, 0.05) = 30.1$. Retain H_0.

(b) $\beta(1.5) = 0.6094$, $\pi(1.5) = 0.3906$.

(c) $\pi(\sigma^2) = 1 - P\left(\chi^2(19) < \frac{30.1}{\sigma^2}\right)$.

5.9　(a) p-value $= 0.0438$. Reject H_0.

(b) p-value $= 0.0847$. Retain H_0.

(c) p-value $= 0.107$. Retain H_0.

5.10　(a) $\chi^2 = 41.413$. Reject H_0 if $\chi^2 = 41.413 \geq \chi^2(29, 0.05) = 42.557$.
Retain H_0.

(b) p-value $= 0.0633$.

(c) $\beta(3) = P(\chi^2(29) < 18.914) = 0.0765 \implies \pi(3) = 0.9235$.

5.11　$\chi^2 = \frac{29(1.8)^2}{2^2} = 23.49$. Reject H_0 if $\chi^2 \leq \chi^2(29, 0.975) = 16.047$ or $\chi^2 \geq \chi^2(29, 0.05/2) = \chi^2(29, 0.025) = 45.722$. Retain H_0.

5.12　(a) Since $\bar{x} = 226.5$, $s^2 = 1.61$, test $H_0 : \sigma^2 = 2.25$ vs. $H_1 : \sigma^2 < 2.25$. The test statistic is $\frac{n-1}{\sigma_0^2} S^2 \sim \chi^2(n-1) \implies \frac{10-1}{2.25} 1.61 = 6.44$. The p-value of the test is $P(\chi^2(9) \leq 6.44) = 0.3047$. Retain H_0. It is plausible that the variance of the thickness is 2.25. Note that the critical region is $(0, 3.325)$, $3.325 = \text{inv}\chi^2(0.95, 9)$.

(b) $\beta(2) = P(\frac{(n-1)S^2}{\sigma_0^2} > 3.325|\sigma^2 = 2) = P(\chi^2(n-1) > 3.325\frac{2.25}{2}|\sigma^2 = 2) = P(\chi^2(n-1) > 3.75) = 0.927$.

5.13　(a) $z = -2.049$. We reject H_0.

(b) p-value $= 0.0405$.

(c) $\beta(0.27) = 0.6434$.

5.14　(a) invNorm$(0.975, 1/6, \sqrt{800(1/6)(5/6)}) = 153.99 \implies x = 154$ or invNorm$(0.025, 1/6, \sqrt{800(1/6)(5/6)}) = 112.67 \implies x = 112$.
Therefore $x \leq 112$ or $x \geq 154$.

(b) $107 \leq x \leq 160$.

5.15　Note that $np_0 = 8(0.6) = 4.8 < 5$ and $n(1 - p_0) = 8(.4) = 3.2 < 5$. The normal approximation is not appropriate in this instance. The binomial distribution must be used directly. Let X be the number of successes.

(a) p-value $= P(X \leq 3) = \text{binomcdf}(8, 0.6, 3) = 0.17367$.

(b) The critical region when $\alpha = 0.1$ is $x \leq 2$ and $x = 8$.

5.16　(a) Reject H_0 if $f = \frac{s_X^2}{s_Y^2} = \frac{20.1^2}{12.2^2} = 2.7144 \geq F(33, 28, 0.025) = 2.089$ or $f \leq F(33, 28, 0.975) = 0.489$. Reject H_0.

(b) Because we rejected H_0 in the test in (a), equal variances cannot be assumed. To test equality of means, the degrees of freedom of the t distribution is given by

$$\nu = \left| \frac{\left(\frac{1}{34} \frac{20.1^2}{12.2^2} + \frac{1}{29} \right)^2}{\frac{1}{34^2(33)} \left(\frac{20.1^2}{12.2^2} \right)^2 + \frac{1}{29^2(28)}} \right| = 55.$$

Reject H_0 if $|t| = \left| \frac{\bar{x} - \bar{y}}{\sqrt{\frac{s_X^2}{m} + \frac{s_Y^2}{n}}} \right| = \left| \frac{105.5 - 90.9}{\sqrt{\frac{20.1^2}{34} + \frac{12.2^2}{29}}} \right| = 3.5395 \geq t(55, 0.025) = 2.004.$

Reject H_0.

(c) p-value $= 2P(t(55) \geq 3.5395) = 2(0.000412) = 0.000824.$

5.17   (a) Reject H_0 if $f = \frac{s_X^2}{s_Y^2} = \frac{6.8}{7.1} = 0.95775 \geq F(10, 9, 0.1) = 2.4163.$ Retain H_0.

(b) Reject H_0 if $f = \frac{s_X^2}{s_Y^2} = \frac{6.8}{7.1} = 0.95775 \leq F(10, 9, 0.95) = 0.3311.$ Retain H_0.

(c) Reject H_0 if $f = \frac{s_X^2}{s_Y^2} = \frac{6.8}{7.1} = 0.95775 \geq F(10, 9, 0.005) = 6.4172$ or $f = 0.95775 \leq F(10, 9, 0.995) = 0.16757.$ Retain H_0.

5.18   (a) Since the variances are assumed equal, we compute the pooled variance as $s_p^2 = 10.186$. Reject H_0 if

$$t = \frac{19.1 - 16.3}{\sqrt{10.186}\sqrt{\frac{1}{24} + \frac{1}{18}}} = 2.8137 \geq t(40, 0.05) = 1.684.$$

Reject H_0.

(b) p-value $= P(t(40) \geq 2.8137) = 0.00378.$

5.19   (a) To test equality of means, the degrees of freedom of the t-distribution is given by $\nu = 37$. Reject H_0 if

$$|t| = \left| \frac{3.8 - 3.6}{\sqrt{\frac{1.2^2}{20} + \frac{1.3^2}{20}}} \right| = 0.50556 \geq t(37, 0.025) = 2.0262.$$

Retain H_0.

(b) p-value $= 2P(t(37) \geq 0.50556) = 2(0.30808) = 0.61616.$

5.20 Reject H_0 if

$$|t| = \left| \frac{(\bar{x} - \bar{y})}{s_p \sqrt{\frac{1}{m} + \frac{1}{n}}} \right| = \frac{|\bar{x} - \bar{y}|}{\sqrt{3581.6} \sqrt{\frac{1}{9} + \frac{1}{14}}} = \frac{|\bar{x} - \bar{y}|}{25.5692} \geq t(21, 0.025) = 2.0796$$

$$\implies |\bar{x} - \bar{y}| \geq 2.0796(25.5692) = 53.1737.$$

The smallest value of $|\bar{x} - \bar{y}|$ resulting in H_0 being rejected is 53.1737.

5.21 (a) The pooled proportion is calculated as $\bar{p}_0 = \frac{500 \cdot \frac{85}{500} + 500 \cdot \frac{93}{500}}{1000} = 0.178$.
Reject H_0 if

$$|z| = \left| \frac{\bar{x} - \bar{y}}{\sqrt{\bar{p}_0(1 - \bar{p}_0)} \sqrt{\frac{1}{n} + \frac{1}{m}}} \right| = \left| \frac{\frac{85}{500} - \frac{93}{500}}{\sqrt{0.178(1 - 0.178)} \sqrt{\frac{1}{500} + \frac{1}{500}}} \right|$$

$$= 0.6614 \geq z_{0.025} = 1.96.$$

Retain H_0.

(b) p-value $= 2P(Z \geq 0.6614) = 2(0.2541) = 0.5082$.

(c) $\beta(-.05) = 0.45743$.

5.22 (a) Let X and Y denote the sample of at bats last and this season, respectively. Consider the hypothesis $H_0 : p_X = p_Y$ vs. $H_1 : p_X > p_Y$ with $\alpha = 0.05$. The pooled proportion is calculated as $\bar{p}_0 = \frac{300(0.276) + 235(0.220)}{535} = 0.2514$. Reject H_0 if

$$z = \frac{0.276 - 0.220}{\sqrt{0.2514(1 - 0.2514)} \sqrt{\frac{1}{300} + \frac{1}{235}}} = 1.4818 \geq z_{0.05} = 1.645.$$

Retain H_0.

(b) p-value $= 0.069$.

5.23 (a) The pooled proportion is calculated as $\bar{p}_0 = \frac{550(0.61) + 690(0.53)}{1240.0} = 0.56548$. Reject H_0 if

$$z = \frac{0.61 - 0.53}{\sqrt{0.56548(1 - 0.56548)} \sqrt{\frac{1}{550} + \frac{1}{690}}} = 2.8234 \geq z_{0.01} = 2.326.$$

Reject H_0.

(b) p-value $= P(Z \geq 2.8234) = 0.002376$.

(c) We show how to do this in general for a two-proportion, one-sided test. In our case $H_1 : p_X - p_Y > 0$. Suppose we have $H_1 : p_X - p_Y = \vartheta > 0$. Then with $SE = \sqrt{\frac{\bar{X}(1-\bar{X})}{n_X} + \frac{\bar{Y}(1-\bar{Y})}{n_Y}}$,

$$\beta(\vartheta) = P\left(\bar{X} - \bar{Y} < z_\alpha \sqrt{\bar{P}_0\left(1 - \bar{P}_0\right)\left(\frac{1}{n_X} + \frac{1}{n_Y}\right)}\right)$$

$$= P\left(\frac{\bar{X} - \bar{Y} - \vartheta}{SE} < \frac{z_\alpha \sqrt{\bar{P}_0\left(1 - \bar{P}_0\right)\left(\frac{1}{n_X} + \frac{1}{n_Y}\right)} - \vartheta}{SE}\right)$$

$$\approx P\left(Z < \frac{z_\alpha \sqrt{\bar{P}_0\left(1 - \bar{P}_0\right)\left(\frac{1}{n_X} + \frac{1}{n_Y}\right)} - \vartheta}{SE}\right)$$

$$\approx P\left(Z < \frac{z_\alpha \sqrt{\bar{p}_0(1 - \bar{p}_0)}\sqrt{\frac{1}{n_X} + \frac{1}{n_Y}} - \vartheta}{\sqrt{\frac{\bar{p}_X(1-\bar{p}_X)}{m} + \frac{\bar{p}_Y(1-\bar{p}_Y)}{n}}}\right)$$

(when samples values are substituted for rvs).

Alternatively, find the critical region for given α first using $x = \text{invNorm}(1 - \alpha, 0, \sqrt{\bar{p}_0(1 - \bar{p}_0)(1/n_X + 1/n_Y)})$. Then find the area to the left of x under the normal curve using normalcdf$(-\infty, x, \vartheta, SE)$.

Given $p_X - p_Y = \vartheta = 0.1$, we have $SE = 0.02817$, and $x = \text{invNorm}(.99, 0, 0.02833) = 0.06591$. Then $\beta(0.1) = \text{normalcdf}(-\infty, 0.06591, 0.1, 0.02817) = 0.11311$.

5.24   (a) $\bar{d} = 0.5, s_D^2 = 7.8333$. Reject H_0 if

$$\frac{0.5 - 0.0}{\frac{\sqrt{7.8333}}{\sqrt{10}}} = 0.56493 \geq t(9, 0.05) = 1.8331.$$

Retain H_0.

(b) p-value $= 0.29296$.

5.25   (a) $\bar{d} = -0.02, s_D^2 = 0.0008222$. Reject H_0 if

$$\frac{-0.02 - 0.0}{\frac{\sqrt{0.0008222}}{\sqrt{10}}} = -2.2057 \geq t(9, 0.025) = 2.2622 \text{ or } -2.2057 \leq -2.2622.$$

Retain H_0.

(b) p-value $= 2(0.027414) = 0.054828$.

5.26   Test $H_0 : p_{O+} = 0.2785, p_{A+} = 0.208, \ldots, p_{AB-} = 0.0049$ vs. H_1: the proportions are not the same. Expected frequencies:

Blood Type	Expected Number of Residents with Blood Type
O+	$1150 * 0.2785 = 320.28$
A+	$1150 * 0.208 = 239.2$
B+	$1150 * 0.3814 = 438.61$
AB+	$1150 * 0.0893 = 102.70$
O-	$1150 * 0.0143 = 16.445$
A-	$1150 * 0.0057 = 6.555$
B-	$1150 * 0.0179 = 20.585$
AB-	$1150 * 0.0049 = 5.635$

Value of $D_7 \sim \chi^2(7)$: $d_7 = 19.934$. Reject H_0 if $d_7 = 19.934 \geq \chi^2(7, 0.01) = 18.475$. Reject H_0.

5.27   Test $H_0 :$ the die is fair vs. $H_1 :$ the die is not fair.

Expected frequencies: $(180)\left(\frac{1}{6}\right) = 30$.

Value of $D_5 \sim \chi^2(5)$: $d_5 = \dfrac{1}{15}(\vartheta - 30)^2$. Reject H_0 if $\frac{1}{15}(\vartheta - 30)^2 \geq \chi^2(5, 0.05) = 11.07 \Rightarrow \vartheta \leq 17.114$ or $\vartheta \geq 42.886$. Reject H_0 if $\vartheta \leq 17$ or $\vartheta \geq 43$.

5.28   Test $H_0 : p_B = 0.207, p_O = 0.205, \ldots, p_B = 0.124$ vs. $H_1 :$ the percentages at the two plants are not the same. Expected frequencies:

Color	Expected Number of M&M
Blue	207
Orange	205
Green	198
Yellow	135
Red	131
Brown	124

Value of $D_5 \sim \chi^2(5)$: $d_5 = 1.8765$. Reject H_0 if $d_5 = 1.8765 \geq \chi^2(5, 0.05) = 11.07$. Retain H_0.

5.29 Test H_0 : data follows a Poisson distribution with $\lambda = 2$ vs. H_1 : data does not follow a Poisson distribution with $\lambda = 2$.

Expected frequencies:

Sailfish Caught	Expected Number of Days
0	$60 * \frac{e^{-2}2^0}{0!} = 8.1201$
1	$60 * \frac{e^{-2}2^1}{1!} = 16.24$
2	$60 * \frac{e^{-2}2^2}{2!} = 16.24$
3	$60 * \frac{e^{-2}2^3}{3!} = 10.827$
≥ 4	$60 * \left(1 - \sum_{k=0}^{3} \frac{e^{-2}2^k}{k!}\right) = 8.5726$

Value of $D_4 \sim \chi^2(4)$: $d_4 = 4.4277$. Reject H_0 if $d_4 = 4.4277 \geq \chi^2(4, 0.05) = 9.4877$. Retain H_0.

5.30 Test H_0 : data follows a Poisson distribution vs. H_1 : data does not follow a Poisson distribution.

(a) The Poisson rate λ can be estimated from 575 observations as

$$\bar{\lambda} = \frac{(229)(0) + (211)(1) + (93)(2) + (35)(3) + (7)(4)}{576} = 0.922.$$

Expected frequencies:

Hummingbird Visits	Expected Number of Days
0	$576 * \frac{e^{-0.922}0.922^0}{0!} = 229.088$
1	$576 * \frac{e^{-0.922}0.922^1}{1!} = 211.219$
2	$576 * \frac{e^{-0.922}0.922^2}{2!} = 97.372$
3	$576 * \frac{e^{-0.922}0.922^3}{3!} = 29.926$
≥ 4	$576 * \left(1 - \sum_{k=0}^{3} \frac{e^{-0.922}0.922^k}{k!}\right) = 8.394$

Value of $D_4 \sim \chi^2(3)$: $d_4 = 1.075$. Reject H_0 if $d_4 = 1.075 \geq \chi^2(3, 0.05) = 7.8147$. Retain H_0.

(b) Now suppose $\lambda = 0.8$. Expected frequencies:

Hummingbird Visits	Expected Number of Days
0	$576 * \frac{e^{-0.8}0.8^0}{0!} = 258.813$
1	$576 * \frac{e^{-0.8}0.8^1}{1!} = 207.051$
2	$576 * \frac{e^{-0.8}0.8^2}{2!} = 82.820$
3	$576 * \frac{e^{-0.8}0.8^3}{3!} = 22.085$
≥ 4	$576 * \left(1 - \sum_{k=0}^{3} \frac{e^{-0.8}0.8^k}{k!}\right) = 5.230$

Value of $D_4 \sim \chi^2(4)$: $d_4 = 13.780$. Reject H_0 if $d_4 = 13.780 \geq \chi^2(4, 0.05) = 9.4877$. Reject H_0.

5.31 H_0 : data follows an exponential distribution with mean 40 seconds vs. H_1 : data does not follow and exponential distribution with mean 40 seconds. Probabilities of an interarrival time falling in each of the intervals are listed in the following tables.

Interval	P (time falling in interval)
$[0, 20)$	$e^{-\frac{1}{40}(0)} - e^{-\frac{1}{40}(20)} = 0.393$
$[20, 40)$	$e^{-\frac{1}{40}(20)} - e^{-\frac{1}{40}(40)} = 0.239$
$[40, 60)$	$e^{-\frac{1}{40}(40)} - e^{-\frac{1}{40}(60)} = 0.145$
$[60, 90)$	$e^{-\frac{1}{40}(60)} - e^{-\frac{1}{40}(90)} = 0.118$
$[90, 120)$	$e^{-\frac{1}{40}(90)} - e^{-\frac{1}{40}(120)} = 0.055$
$[120, 180)$	$e^{-\frac{1}{40}(120)} - e^{-\frac{1}{40}(180)} = 0.039$
$[180, \infty)$	$e^{-\frac{9}{2}} = 0.011$

Interval	Expected No. Interarrival Times in Interval
$[0, 20)$	39.3
$[20, 40)$	23.9
$[40, 60)$	14.5
$[60, 90)$	11.8
$[90, 120)$	5.5
$[120, \infty)$	5.0

Value of $D_5 \sim \chi^2(5)$: $d_5 = 5.3826$. Reject H_0 if $d_5 = 5.3826 \geq \chi^2(5, 0.05) = 11.07$. Retain H_0. The possibility that the data is from an exponential distribution with $\lambda = \frac{1}{40}$ cannot be eliminated.

5.32 H_0 : data follows a geometric distribution vs. H_1 : data does not follow a geometric distribution. If X is the random variable that counts the number of casts, and if X is geometric, then $P(X = x) = (1 - p)^{x-1} p$. We have the estimate for p as $\bar{p} = 1/3.82 = 0.26178$.

Expected frequencies:

Number of Casts Until a Strike	Expected Frequency
1	$(50)(0.26178) = 13.089$
2	$(50)(1 - 0.26178)(0.26178) = 9.6626$
3	$(50)(1 - 0.26178)^2(0.26178) = 7.1331$
4	$(50)(1 - 0.26178)^3(0.26178) = 5.2658$
5	$(50)\sum_{i=4}^{\infty}(1 - 0.26178)^i(0.26178) = 14.8495$

We combine the cells for $i \geq 5$ strikes so that all the expected frequencies are at least 5. Value of $D_4 \sim \chi^2(3)$: $d_4 = 9.277$. Reject H_0 if $d_4 = 9.277 \geq \chi^2(3, 0.01) = 11.345$. Retain H_0. However, if $\alpha = 0.05$, the null hypothesis can be rejected since

$$d_4 = 9.277 \geq \chi^2(3, 0.05) = 7.8147.$$

At that level, it is likely that the data is not following a geometric distribution.

5.33 Test H_0 : injury due to criminal violence and choice of profession are independent vs. H_1 : injury due to criminal violence and choice of profession are not related.

Estimated row and column probabilities:

$$\hat{p}_{V*} = \frac{318}{490} = 0.64898, \, \hat{p}_{O*} = \frac{172}{490} = 0.35102, \, \hat{p}_{*P} = \frac{174}{490} = 0.3551,$$

$$\hat{p}_{*C} = \frac{116}{490} = 0.23673, \, \hat{p}_{*T} = \frac{99}{490} = 0.20204, \, \hat{p}_{*S} = \frac{101}{490} = 0.20612.$$

Estimated frequencies:

$$490\hat{p}_{V*}\hat{p}_{*P} = 490 \cdot \frac{318}{490} \cdot \frac{174}{490} = 112.922, \quad 490\hat{p}_{V*}\hat{p}_{*C} = 490 \cdot \frac{318}{490} \cdot \frac{116}{490} = 75.280,$$

$$490\hat{p}_{V*}\hat{p}_{*T} = 490 \cdot \frac{318}{490} \cdot \frac{99}{490} = 64.249, \quad 490\hat{p}_{V*}\hat{p}_{*S} = 490 \cdot \frac{318}{490} \cdot \frac{101}{490} = 65.546,$$

$$490\hat{p}_{O*}\hat{p}_{*P} = 490 \cdot \frac{172}{490} \cdot \frac{174}{490} = 61.077, \quad 490\hat{p}_{O*}\hat{p}_{*C} = 490 \cdot \frac{172}{490} \cdot \frac{116}{490} = 40.718,$$

$$490\hat{p}_{O*}\hat{p}_{*T} = 490 \cdot \frac{172}{490} \cdot \frac{99}{490} = 34.751, \quad 490\hat{p}_{O*}\hat{p}_{*S} = 490 \cdot \frac{172}{490} \cdot \frac{101}{490} = 35.453.$$

Value of $D_7 \sim \chi^2(3)$: $d_7 = 65.526$. Reject H_0 if $d_7 = 65.526 \geq \chi^2(3, 0.01) = 11.345$. Reject H_0.

5.34 Test H_0 : selectivity and sensitivity are independent vs. H_1 : selectivity and sensitivity are not independent

Estimated row and column probabilities:

$$\hat{p}_{LS*} = \frac{30}{170} = 0.176\,47,\ \hat{p}_{AS*} = \frac{112}{170} = 0.658\,82,\ \hat{p}_{HS*} = \frac{28}{170} = 0.164\,71,$$

$$\hat{p}_{*LN} = \frac{52}{170} = 0.305\,88,\ \hat{p}_{*AN} = \frac{88}{170} = 0.517\,65,\ \hat{p}_{*HN} = \frac{30}{170} = 0.176\,47.$$

Estimated frequencies: (Note that all estimated frequencies are at least 5 except the frequency of high selectivity and high sensitivity. However, it is acceptably close to 5.)

$$170\hat{p}_{LS*}\hat{p}_{*LN} = 170 \cdot \frac{30}{170} \cdot \frac{52}{170} = 9.176,\ 170\hat{p}_{LS*}\hat{p}_{*AN} = 170 \cdot \frac{30}{170} \cdot \frac{88}{170} = 15.529,$$

$$170\hat{p}_{LS*}\hat{p}_{*HN} = 170 \cdot \frac{30}{170} \cdot \frac{30}{170} = 5.294,\ 170\hat{p}_{AS*}\hat{p}_{*LN} = 170 \cdot \frac{112}{170} \cdot \frac{52}{170} = 34.258,$$

$$170\hat{p}_{AS*}\hat{p}_{*AN} = 170 \cdot \frac{112}{170} \cdot \frac{88}{170} = 57.976,\ 170\hat{p}_{AS*}\hat{p}_{*HN} = 170 \cdot \frac{112}{170} \cdot \frac{30}{170} = 19.765,$$

$$170\hat{p}_{HS*}\hat{p}_{*LN} = 170 \cdot \frac{28}{170} \cdot \frac{52}{170} = 8.565,\ 170\hat{p}_{HS*}\hat{p}_{*AN} = 170 \cdot \frac{28}{170} \cdot \frac{88}{170} = 14.495,$$

$$170\hat{p}_{HS*}\hat{p}_{*HN} = 170 \cdot \frac{28}{170} \cdot \frac{30}{170} = 4.9411.$$

Value of $D_8 \sim \chi^2(4)$: $d_8 = 18.012$. Reject H_0 if $d_8 = 18.012 \geq \chi^2(4, 0.01) = 13.277$. Reject H_0.

5.35 Test H_0 : rH factor and blood type are independent traits vs. H_1 : rH factor and blood type are not independent.

Reformulate the table in terms of rH factor.

	O	A	B	AB	Totals
Positive rH	344	207	448	92	1091
Negative rH	23	12	23	11	69
Totals	367	219	471	103	1150

Value of $D_7 \sim \chi^2(3)$: $d_7 = 5.2709$. Reject H_0 if $d_7 = 5.270\,9 \geq \chi^2(3, 0.05) = 7.814\,7$. Retain H_0.

5.36 Test H_0 : gender and genre of programming are independent vs. H_1 : gender and genre of programming are dependent. Value of $D_5 \sim \chi^2(2)$: $d_5 = 4.4334$. Reject H_0 if $d_5 = 4.4334 \geq \chi^2(2, 0.05) = 5.9915$. Retain H_0.

5.37 Test $H_0 : p_I = p_{II} = p_{III} = p_{IV} = p_V$ vs. H_1 : the proportions are not the same.

Estimated row and column probabilities:

$$\hat{p}_{K*} = \frac{112}{500} = 0.224, \, \hat{p}_{N*} = \frac{388}{500} = 0.776, \, \hat{p}_{*i \in \{I,II,III,IV,V\}} = \frac{100}{500} = 0.2.$$

Estimated frequencies:

$$500 \hat{p}_{K*} \hat{p}_{*i \in \{I,II,III,IV,V\}} = 500 \cdot \frac{112}{500} \cdot \frac{100}{500} = 22.4,$$

$$500 \hat{p}_{N*} \hat{p}_{*i \in \{I,II,III,IV,V\}} = 500 \cdot \frac{388}{500} \cdot \frac{100}{500} = 77.6.$$

Value of $D_9 \sim \chi^2(4)$: $d_9 = 2.140$. Reject H_0 if $d_9 = 2.140 \geq \chi^2(4, 0.05) = 9.4877$. Retain H_0.

5.38 Test H_0 : A and B traits are independent vs. H_1 : the A and B traits are dependent.

Estimated row and column probabilities:

$$\hat{p}_{A_i *} = \frac{100}{300}, \, i = 1, 2, 3, \qquad \hat{p}_{*B_j} = \frac{75}{300}, \, j = 1, 2, 3, 4.$$

Estimated frequencies:

$$300 \hat{p}_{A_i *} \hat{p}_{*B_j} = 300 \cdot \frac{100}{300} \cdot \frac{75}{300} = 25, \, i = 1, 2, 3, \, j = 1, 2, 3, 4.$$

Value of $D_{11} \sim \chi^2(6)$: $d_{11} = \frac{104}{25} \vartheta^2 = \chi^2(6, 0.05) = 12.59 \implies \vartheta = 1.74$. The smallest value of ϑ for which the null hypothesis is rejected is $\vartheta = 2$.

5.39 Let S denote the number of speeds greater than 200 mph. From the table, $S = 8$. Set $\alpha = 0.05$. We choose the minimum value of k such that $\sum_{i=1}^{k} \binom{10}{i}(0.5)^{10} \geq 0.95$. The relevant section of the binomial table for $n = 10$ and $p = 0.5$ is given below.

k	$P(S \leq k \mid H_0$ is true)
6	0.827 15
7	0.944 34
8	0.988 28
9	0.998 05

We choose $k = 8$. We reject H_0 if $S > 8$. Since $S = 8$, we retain H_0. The p-value is $P(S \geq 8) = \sum_{i=8}^{10} \binom{10}{i}(0.5)^{10} = 0.05468$.

5.40 $H_0 : \mu_A = \mu_B = \mu_C$ vs. H_1 : means are different.

(a) $\bar{x}_{**} = 9.6717,\quad \bar{x}_{*A} = 10.968,\quad \bar{x}_{*B} = 10.68,\quad \bar{x}_{*C} = 7.3667.$

(b) $SSE = 52.61,\quad SSTR = 48.061,\quad SST = SSE + SSTR = 100.67.$

(c) $MSTR = \frac{SSTR}{2} = \frac{48.061}{2} = 24.031, MSE = \frac{SSE}{15} = \frac{52.61}{15} = 3.5073.$

(d) $f = \frac{MSTR}{MSE} = \frac{24.031}{3.5073} = 6.8517.$

(e) p-value $= P(F(2, 15) \geq 6.8517) = 0.008.$

(f)

Source	DF	SS	MSS	F-statistic	p-Value
Treatment	2	48.061	24.031	$f = 6.8517$	0.008
Error	15	52.61	3.5073	*	*
Total	17	100.67	*	*	*

(g) Reject H_0 at the $\alpha = 0.01$ level. The three random variables A, B, and C do not have the same mean.

5.41 $H_0 : \mu_A = \mu_J = \mu_B$ vs. H_1 : the means are not equal.

(a) $\bar{x}_{**} = 9.4,\quad \bar{x}_{*A} = 9.6,\quad \bar{x}_{*J} = 8.8,\quad \bar{x}_{*B} = 9.8.$

(b) $SSE = 10.8,\quad SSTR = 2.8,\quad SST = SSE + SSTR = 10.8 + 2.8 = 13.6.$

(c) $MSTR = \frac{SSTR}{2} = \frac{2.8}{2} = 1.4, MSE = \frac{SSE}{12} = \frac{10.8}{12} = 0.9.$

(d) $f = \frac{MSTR}{MSE} = \frac{1.4}{0.9} = 1.5556.$

(e) p-value $= P(F(2, 12) \geq 1.5556) = 0.2508.$

(f)

Source	DF	SS	MSS	F-statistic	p-Value
Treatment	2	2.8	1.4	$f = 1.5556$	0.2508
Error	12	10.8	0.9	*	*
Total	14	13.6	*	*	*

(g) Retain H_0 at the $\alpha = 0.05$ level. Alice, Bob, and John appear to have the same performance level on installing windshields.

5.42 $H_0 : \mu_P = \mu_L = \mu_N$ vs. H_1 : means are not equal.

Source	DF	SS	MSS	F-statistic	p-Value
Treatment	2	1484.933	742.4667	$f = 11.26657$	.001761
Error	12	790.8	65.9	*	*
Total	14	2275.7	*	*	*

Reject H_0 at the $\alpha = 0.01$ level. Dosage levels seem to have an effect on depression.

5.43 $H_0 : \mu_M = \mu_S = \mu_L$ vs. H_1 : the means are not equal.

Source	DF	SS	MSS	F-statistic	p-Value
Treatment	2	10.0882	5.0441	$f = 2.76338$	.099985
Error	13	23.7293	1.8253	*	*
Total	15	33.818	*	*	*

We retain H_0 at the $\alpha = 0.05$ level. It appears the dairies have the same level of Strontium-90 contamination in their milk.

5.44 The solution is broken up into several steps.

(a) $df_{SSTR} = 3$.

(b) $f = \frac{0.708}{0.75} = 0.9444$.

(c) $SSE = (20)(0.9444) = 18.88$.

(d) $n - 1 = 23 \Rightarrow n = 24$ and $n - k = 20 \Rightarrow 24 - k = 20 \Rightarrow k = 4$.

(e) p-value= $P(F(3, 20) \geq 0.75) = 0.5351$.

(f) Completed table:

Source	DF	SS	MSS	F-statistic	p-Value
Treatment	**3**	2.124	0.708	$f = 0.75$	**0.5351**
Error	20	**18.88**	**9.44**	*	*
Total	**23**	**21.004**	*	*	*

5.45 $H_0 : \mu_{5Y} = \mu_{10Y} = \mu_{15Y} = \mu_{20Y}$ vs. H_1 : the means are different.

(a) $\bar{x}_{**} = 50.35$, $\bar{x}_{*5Y} = 58.4$, $\bar{x}_{*10Y} = 57.4$, $\bar{x}_{*15Y} = 43.6$, $\bar{x}_{*20Y} = 42.0$.

(b) $SSE = 201.6$, $SSTR = 1149.0$, $SST = SSE + SSTR = 1350.6$.

(c) $MSTR = \frac{SSTR}{3} = \frac{1149.0}{3} = 383.0$, $MSE = \frac{SSE}{16} = \frac{201.6}{16} = 12.6$.

(d) $f = 30.397$.

(e) p-value= $P(F(3, 16) \geq 30.397) = 7.6621 \times 10^{-7}$.

(f)

Source	DF	SS	MSS	F-statistic	p-Value
Treatment	3	1149.0	383.0	$f = 30.397$	7.6621×10^{-7}
Error	16	201.6	12.6	*	*
Total	19	1350.6	*	*	*

(g) Reject H_0 at the $\alpha = 0.01$ level. Marksmanship skill differs according to how many years served.

A.6   ANSWERS TO CHAPTER 6 PROBLEMS

6.2 $(x - \bar{x}) = r\frac{s_X}{s_Y}(y - \bar{y})$. $\hat{b} = r\frac{s_X}{s_Y}, \hat{a} = \bar{x} - \hat{b}\bar{y}$. Finally, $f(\hat{a}, \hat{b}) = (n-1)(1-r^2)s_X^2$.

6.3   We have $b = \rho\frac{\sigma_Y}{\sigma_X}$, $d = \rho\frac{\sigma_X}{\sigma_Y} \implies b \times d = \rho^2$. Then using the two lines given, $\rho = 0.7027$.

6.4   (a) $Y_1 = 8.6 + \varepsilon_1, Y_2 = 11.1 + \varepsilon_2 \implies Y_1 - Y_2 \sim N(-2.5, 1.7\sqrt{2})$.

(b) $P(Y_1 > Y_2) = \text{normalcdf}(0, \infty, -2.5, 1.7\sqrt{2}) = 0.1492$.

6.5   Since $\bar{x} = 4.45, s_x = 2.167, \bar{y} = 5.375, s_y = 1.2104, r = 0.7907$ we have

$$(y - 5.375) = 0.7907\frac{1.2104}{2.167}(x - 4.45) = 0.4416(x - 4.45)$$

and $(x - 4.45) = 0.7907\dfrac{2.167}{1.2104}(y - 5.375) = 1.4156(y - 5.375)$. The minimum of f with dependent variable x is $7(1 - 0.7907^2)1.2104^2 = 3.844$, and with dependent variable y is $7\sqrt{1 - 0.7907^2}2.167^2 = 12.32$.

6.6   (a) We have $VSAT - 610 = 0.73\frac{120}{110}(MSAT - 570)$. Therefore, if $MSAT = 690, VSAT = 705.56$.

(b) $MSAT = 570 + 0.73(110/120)(700 - 610) = 630.23$.

(c) $0.738846 = (MSAT - 570)/110 = 0.73(VSAT - 610)/120 \implies z = (VSAT - 610)/120 = 1.01211 \implies \text{normalcdf}(-\infty, z) = 0.842$.

(d) $S_e = \sqrt{1 - r^2}SD_{VSAT} = 82.0136$.

6.7   (a) We have the model $S \sim N(a + bB, \sigma)$ and $\sigma \approx s_e = \sqrt{31/30}SD_S\sqrt{1 - 0.26^2} = 2.159$. So, $\text{normalcdf}(68, \infty, 62, 2.2.159) = 0.00273$.

(b) The regression equation is $S = 45.7933 + 0.2383 B$. If $B = 72$, then $S = 62.9509$ and then $\sim N(62.95, s_e)$, so $\text{normalcdf}(68, \infty, 62.9509, 2.1259) = 0.00967$.

6.8   (a) 16.8.   (b) 15.2.

6.9   (a) 19%   (b) 63.20%   (c) 52.50%   (d) 50%

6.10   Consider $f(b) = E(Y - bX)^2$. We have $f'(b) = -2E(Y - bX)X = 0 \implies b = E(XY)/E(X^2)$. Also $f''(b) = 2EX^2 > 0$. Using observations we have, $\hat{b} = \frac{\sum x_i y_i}{\sum x_i^2}$.

6.11   In matrix form $\begin{pmatrix} n & \sum x_i & \sum x_i^2 \\ \sum x_i & \sum x_i^2 & \sum x_i^3 \\ \sum x_i^2 & \sum x_i^3 & \sum x_i^4 \end{pmatrix}\begin{pmatrix} a \\ b \\ c \end{pmatrix} = \begin{pmatrix} \sum y_i \\ \sum x_i y_i \\ \sum x_i^2 y_i \end{pmatrix}$. For the given data, the equation of the least squares quadratic is $y = 0.57 + x + 1.107x^2$.

6.12 $E(S^2/\sigma^2) = n - 2$ so an unbiased estimator for σ^2 using the s^2 of the problem is $\frac{n}{n-2}s^2$.

6.14   Let x denote gas CPI and y be the food CPI. The data gives $\bar{x} = 4.836, \bar{y} = 2.182, s_X = 26.95, s_Y = 1.88$. The regression equation is food CPI $= 2.323 - 0.0293$ gas CPI. The correlation coefficient is $r = -0.42$.

6.17   Here is the ANOVA table

Source of Variation	DF	SS	MS	F-statistic	p-Value
Regression	1	570.04	$MSR = 570.04$	$77.05 = F(1, 18)$	0.0
Residuals	18	133.16	$MSE = 7.4$		
Total	19	703.2			

The last column gives the p-value for the observed F value of 77.05. The null is rejected and there is strong evidence that the slope of the regression line is not zero. Since $r\frac{s_y}{s_x}$ is the slope, we can use this to conclude that there is strong evidence the correlation is not zero.

6.18   We get the regression line and correlation coefficient $y = -3.10091 + 2.02656x, r = .99641$. We want to test $H_0 : \beta = 0, H_1 : \beta \neq 0$. We get the test statistic $t = 33.2917$. With $n - 2 = 8$ degrees of freedom, the p-value is basically 0.

6.24   We have $\hat{y}(750) = 14.95$ and the 95% PI 14.95 ± 1.492.

6.25   We have $S_{yy} = 2243266 - \frac{5712^2}{17} = 324034$ and

$$S_{xx} = 35990 - \frac{660^2}{17} = 10366.5, S_{xy} = 188429 - \frac{(660 \times 5712)}{17} = -33331.$$

Then $\hat{b} = \frac{S_{xy}}{S_{xx}} = -3.22$ and $\hat{\alpha} = \bar{y} - \hat{b}\bar{x} = 461.01$. Finally, $s^2 = \frac{1}{n-2}(S_{yy} - \frac{S_{xy}^2}{S_{xx}}) = 14457.7$.

Authors' Biographies

EMMANUEL BARRON

Professor Barron received his B.S. (1970) in Mathematics from the University of Illionois at Chicago and his M.S. (1972) and Ph.D. (1974) from Northwestern University in Mathematics specializing in partial differential equations and differential games. After receiving his Ph.D., Dr. Barron was an Assistant Professor at Georgia Tech, and then became a Member of Technical Staff at Bell Laboratories. In 1980 he joined the Department of Mathematical Sciences at Loyola University Chicago, where he is a Professor of Mathematics and Statistics. Professor Barron has published over 70 research papers, and he has also authored the book *Game Theory: An Introduction*, 2nd Edition in 2013. Professor Barron has received continuous research funding from the National Science Foundation and the Air Force Office for Scientific Research. Dr. Barron has taught Probability and Statistics to undergraduates and graduate students since 1974.

JOHN DEL GRECO

A native of Cleveland, Ohio, Dr. John G. Del Greco holds a B.S. in Mathematics from John Carroll University, an M.A. in Mathematics from the University of Massachusetts, and a Ph.D. in Industrial Engineering from Purdue University. Before joining Loyola's faculty in 1987, Dr. Del Greco worked as a systems analyst for Micro Data Base Systems, Inc., located in Lafayette, Indiana. His research interests include applied graph theory, operations research, network flows, and parallel algorithms, and his publications have appeared in such journals as *Discrete Mathematics, Discrete Applied Mathematics, the Computer Journal, Lecture Notes in Computer Science,* and *Algorithmica.* He has been teaching Probability and Statistics at all levels in the Department of Mathematics and Statistics at Loyola for the past 20 years.

Index